AF595568

A DIFFERENT ENERGY

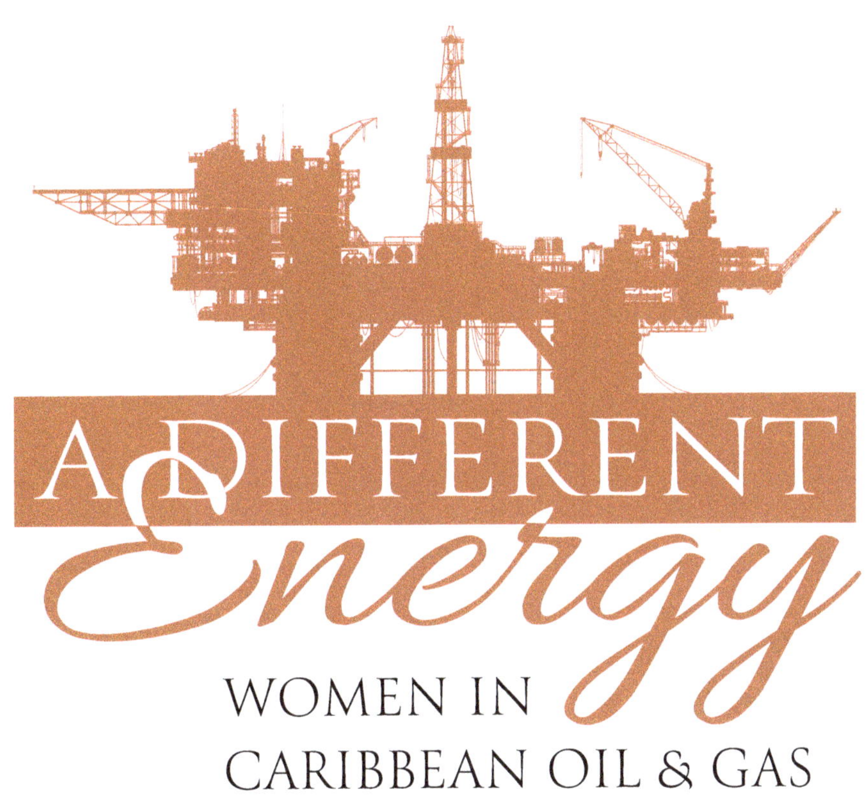

Celeste Mohammed

A member of the United Nations Global Compact Initiative

www.wordsmattercommunications.org

Design and layout by Paria Publishing Company Limited
www.pariapublishing.com

ISBN: 978-976-8291-94-3
Typeset in Canto and Allura
Printed by Scrip J

"...if you want anything said, ask a man.

If you want anything done, ask a woman.."

Margaret Thatcher

"Be courageous, be authentic. You do not have to mimic what is expected. Bring the *you* into what you do and that is passion, that is unique."

Mala Baliraj
Chief Executive Officer, Massy/Wood

Contents

Acknowledgements

Excepting one, I did not know any of the women featured in this book. Yet, they answered my calls, responded to my emails, and made time for hours-long chats – in some cases, inviting me into their homes. They shared their stories generously and trusted me, even when the shock of seeing their own lives in black-and-white induced trembling feet. So, to Arlene, Giselle, Deborah, Satira, Marny, Vandana, Sharista and Grace, I say a heartfelt and humble thank you.

To the other women – Mala, Anne, Mechelle, Natasha, Michelle – who readily shared their wisdom in the form of quotes, thanks to you as well.

One of this book's most vocal advocates has been Garth Chatoor. I also drew much support and encouragement from the erudite minds of Andrew Jupiter, Ellis Lewis and Zaid Khan. Thank you to Aunty Ava for your willingness to "bat for me"; thank you, dearest Melanie for praying and holding my hand through this process.

Thank you, Alice Besson and Denise Mohammed, for effortless working relationships which still managed to produce output that meets my (sometimes) unreasonably high standards.

Bless you, Debbie.

Thank you, Lord.

About the Author

Celeste Mohammed is a Trinidadian lawyer-turned-writer and the author of "Pleasantview"(Ig, Jacaranda, 2021; Ouida, 2022) which won the 2022 OCM Bocas Prize for Caribbean Literature, the 2022 CLMP Firecracker Award for Fiction, and was a finalist for the UK Society of Authors McKitterick Prize for Fiction.

Celeste holds an MFA in Creative Writing from Lesley University, Cambridge, Massachusetts. Her short stories have won numerous awards including a 2018 PEN/Robert J. Dau Short Story Prize.

Celeste's goal is to dispel all myths about island life and island people and to highlight the various points of intersection between the Caribbean and the wider world.

A Different Energy is Celeste's Non-Fiction debut.

"Ultimately to hold a job in the oil and gas industry a woman must learn to balance her responsibilities at home with a very time consuming career, while most senior, male oil executives have a wife."

Anne Ghent
Former Chief Executive Officer,
Ventrin Petroleum Company Limited

Trinidad and Tobago

CHAPTER 1

Her-story of Oil & Gas in Trinidad

IL IS TO THE ISLAND OF TRINIDAD what blood is to the body of a man – or a woman. The first oil well in the Western Hemisphere was drilled in Trinidad, in 1857, by the Merrimac Company.[1] Although that well is often regarded as the start of the Caribbean oil industry in the modern age, one might say Trinidad's petrochemical exports began centuries earlier, in 1595, when the English pirate-explorer Sir Walter Raleigh came looking for El Dorado – the fabled city of gold – and was shown instead, by the Amerindians, the "black gold" of bitumen or *piche* at Tierra de Brea (now La Brea Village); he used the stuff for caulking his ships. He found our oily deposits to be "most excellent good, and melteth not with the sunne as the pitch of Norway, and therefore for ships trading the south partes very profitable."[2]

Very profitable indeed. But for whom? Everywhere oil is discovered, that's always the question. Who should profit? To whose benefit should this windfall redound? The foreign multinational oil companies, the government? The expatriates, the locals? Not many of us, though, question whether the benefits should flow more to men or to women, because we assume that oil should and does bestow wealth equally, regardless of gender. But that's not true. Oil and gas is one of the largest, most lucrative, and most politically powerful industries in the world, yet it employs very few women, tends to lose the women it does hire,[3] and in developing countries, it disproportionately burdens women.[4]

Sir Walter Raleigh caulking his ship with pitch from the Pitch Lake.

Ask anybody working in the petrochemical industry and they will tell you, "it's a man's world". Academics have used more nuanced phraseology, referring to the "profoundly gendered nature of oil work."[5] I saw this with my own eyes while growing up in San Fernando, "the industrial capital" of Trinidad and Tobago, during the 1980s–1990s. Throughout my secondary school years, I had many friends who lived "on camp", one or other of the oil company residential camps along the southern oil belt of Trinidad. Then there were other friends, who had a parent "working in the oil" but didn't live on camp – although their family still got to drive past security and go up to the club and play at the pool. But, in either case, the parent whose employment allowed access to oil's privileges, was always a man. Nobody's "Mummy" was ever the one "working in the oil" or on senior staff. It was always the father.

Male hegemony in the oil and gas industry was accepted as normal back then, when I was growing up – and I'm not that old – and we didn't even wonder, *Where are the women?* We just accepted that it had to be so. And here's another interesting irony: we – my contemporaries and I – constitute The Notorious OBG, the "oil-boom generation" who was born during Trinidad's oil boom of the 1970s, but grew up in the 1980s during the deepest economic recession in Trinidadian history, and as teenagers experienced the resulting attempted coup of 1990. Trauma, upon trauma, upon trauma. The message we received was that oil is a kind of liquid Zeus: a fickle god who gives generously to male providers, and then "raffs" away capriciously, leaving their women either scrambling to make ends meet at home, or exiled in America cleaning toilets and babysitting other people's pickney.

Now, as the mother of a daughter, living through this crucial Decade of Action where the UN has set a 2030 deadline for the achievement of gender equality and the empowerment of all women and girls,[6] I do wonder how many OBG girl-children, like myself, might have aspired differently and made different life choices if we had known that, all this time, there were indeed women succeeding in the oil and gas industry.

I know better now. So, I thought I would ask around for them, ask until I found some such, and then ask them what it's been like to be invisible for so long. That is the purpose of this book. To do exactly what the rapso music of my teenaged years exhorted me to do:

So, Boom Generation!
This power is your own.
Don't stand up and wonder –
Time to claim the throne.
This is the oil-spill session
So free up your heart
Just throw away the water
Cause the fire done start.[7]

I am not a social scientist nor academic, so there's no methodology to disclose, no empirical evidence to release – this ain't that kinda book. However, as a professional woman myself – a corporate lawyer – who has endured the scorching discomfort of being the only female trying to lean in at a boardroom table of males; and who is also a writer and connoisseur of good narrative, I went seeking the undiluted stories of Caribbean oil-women and the wisdom of how they survived having their feet held to the fire.

It won't surprise you, however, that given the male domination of the industry, I found myself beginning by talking to a man: Professor Andrew Jupiter, a Distinguished Fellow of the University of the West Indies, St Augustine, Trinidad. It is mostly men[8] – Prof. Jupiter[9], Ellis Lewis[10], Gérard Besson[11] –

who have been chronicling our oil history, and understandably, as men, they have written about the struggles and triumphs of oilmen and about matters that touch and concern men. Their service has been invaluable, and if asked to recommend essential resources on the establishment and development of the petrochemical industry in Trinidad and Tobago, I would wholeheartedly recommend their tomes.

Early well in the deep forest of southern Trinidad.

But there is a story that has remained largely missing from their accounts – the "her-story", if you will. For, while scouring their writings, I discovered a line here and a line there which proved that women have been involved in the oil industry for as long as men have – just not in as great numbers. Prof. Jupiter's own description of the events surrounding Trinidad's first commercial oil well, attests to the fact that we were there, from the beginning:

> "It happened in 1908, in Point Fortin village, and there was confusion when the residents around saw the explosion. They said, 'This is an act of God!' Oil flew in the air hundreds of metres because they didn't have equipment to control the pressure, and nobody in the village had seen that before. When others heard what was happening in Point Fortin, people flooded from all over Trinidad and the Caribbean and, as a result of that very increase, that's when men left their families and came. Some women came, too. My mother was among them, she came from the far North, from Grande Riviere, to Point Fortin because her brothers found work in the oil."[12]

Walking the tracks to one of the remote oil fields in the forest.

But what did the women come to do?

The earliest period of oil exploration in Trinidad, from 1860s to 1930s, was the era of "wildcat drilling" in heavily forested, jungle areas in the deep South of the island.[13] Drilling was dirty, physical, dangerous, and not dissimilar to the pirate-explorer work of Sir Raleigh in his quest for gold. According to Ellis Lewis[14], Trinidad was described as "the graveyard of geologists" and it wasn't uncommon, at drilling camps, for there to be an outbreak of malaria or for vampire bats to attack workers in their sleep. "Women were neither suited to, nor welcome or accommodated at, these drilling camps," says Prof. Jupiter.[15]

And yet, Besson writes that they came, even into the forest to work alongside men.[16]

Quoting PET O'Connor, a Trinidadian employee of Kern (Trinidad) Oilfields at the time, Besson explains that in those early days, oil was pumped out of the well and flowed into large pits which surrounded the site, and there were few labour-saving devices. "Generally, the thick forest around a well was cleared by hand with axes, and land and roadways graded with shovels and forks by men (and women!) who were called the 'Tattoo Gang' in local parlance...the miles of road through our forests remain as a silent monument to the very special breed of men and their women-folk who have passed into oblivion!"

In the early 20th century, despite the rapid modernisation of the petroleum industry in Trinidad,[17] the oilfields themselves were still considered "fever holes from which few returned alive". According to Prof. Jupiter, a kind of paternalism set in, and rules were passed to the effect that women were not allowed on site.

While drilling for oil was one part of the work, refining it was the other. Trinidad's refineries – at Point Fortin, Pointe-à-Pierre, and later Santa Flora – drew their labour force primarily from the male population of nearby villages. But that's not to say women had no role. As Prof. Jupiter recalls in relation to Point Fortin:

> "I cannot forget the picture at lunchtime, at 11:30. The entire village would hear the Shell horn, they would blow it for lunch. And you will see at that point in time, ladies walking to the gate bringing lunch to their husbands. There were some husbands who would jump on the bike and go home, but so many women flocked to the gate to bring their men lunch...that's as far as they could go...talking to him at the gate. But the contribution of these women was also important. The men worked very hard. They needed to be sustained by good food and so these women were ensuring that the quality of work done was right, as a result of their support and obviously the food."[18]

The oilmen could not be fully productive in a vacuum of maleness. They needed the unglamorous and unheralded – but essential – support of women as domestic workers, family stabilisers, and community anchors.[19]

Still, as Ellis Lewis[20] confirmed, the women could not get past the gates! Except in certain subordinate capacities. The case of the UBOT/Shell refinery at Point Fortin is typical. Picturesque Clifton Hill estate was set aside as a camp for the accommodation of expatriate (read "foreign and white") staff only. There was a club house, playing field for football and cricket, swimming pool, tennis courts, golf course, and even a school for the children of expatriate staff.

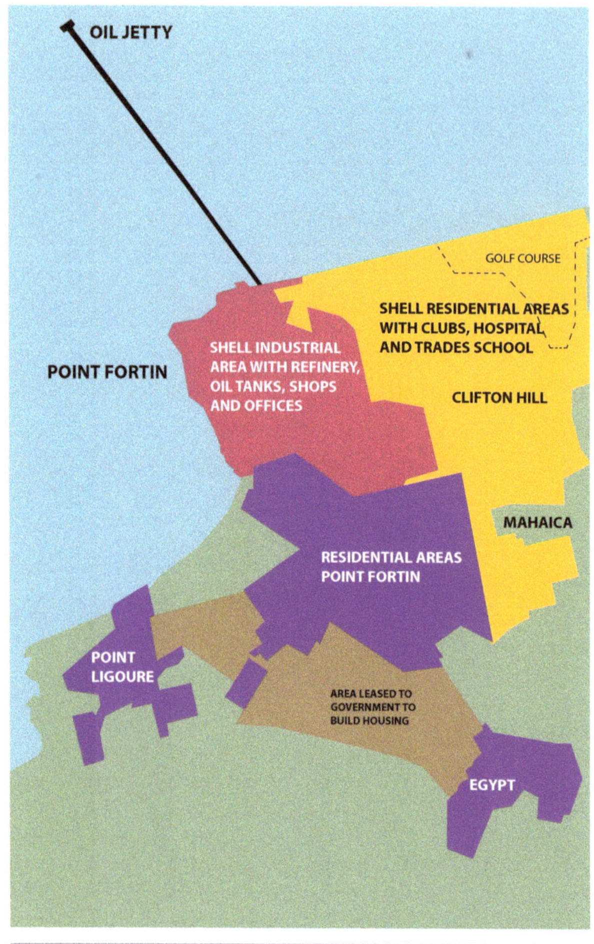

Residential areas around the Point Fortin refinery.

Prof. Jupiter says his mother worked as a housekeeper for a family of expats in Clifton Hill. Mrs Grace-Anne Bradshaw, who would later become headmistress of the Company school, recalls that the teaching staff comprised mainly women sent from England.[21] She also explained the interesting phenomenon of "the expatriate wife": sometimes, an educated woman who would have accompanied her husband who came to work at the oil company, but would not have been catered for in terms of work permits, so she would have kept herself busy by volunteering at the school. Like me, Mrs Bradshaw does not recall any pupil of the school being the child of a Company staff member who was female. Managerial staff at the oil company was always male.

It appears, though, at least from one of Besson's sources, that throughout the turbulent years of the 1930s and 1940s, local women – unpropertied and disenfranchised as they might have been – asserted themselves within the industry's labour movement.

By the 1930s, while the expatriates were living a life of luxury at Clifton Hill, the housing and living conditions in the local oil communities, "the shanty towns on the perimeter of the fields", were so rough that they became what the oil companies viewed as "hotbeds of hooliganism". According to one commentator, "as [is] often the case with people of primitive education, the women were mainly responsible for some of the very regrettable incidents which took place..."[22]

In preparation for the 1937 oilfield strikes, the infamous labour organiser, Tubal Uriah "Buzz" Butler, gathered a Women's Committee to prepare meals and deliver them to the strikers, but some women got more involved than that. They marched and picketed alongside the men. For her part in the Point Fortin riots, Elma Francois became the first woman in Trinidad and Tobago to be charged and tried for sedition. She stood alongside her male counterparts, like Butler, in what came to be known as the Sedition Trials of 1937–1938. In a show of true strength, she represented herself in court and was found not guilty by the jury.[23]

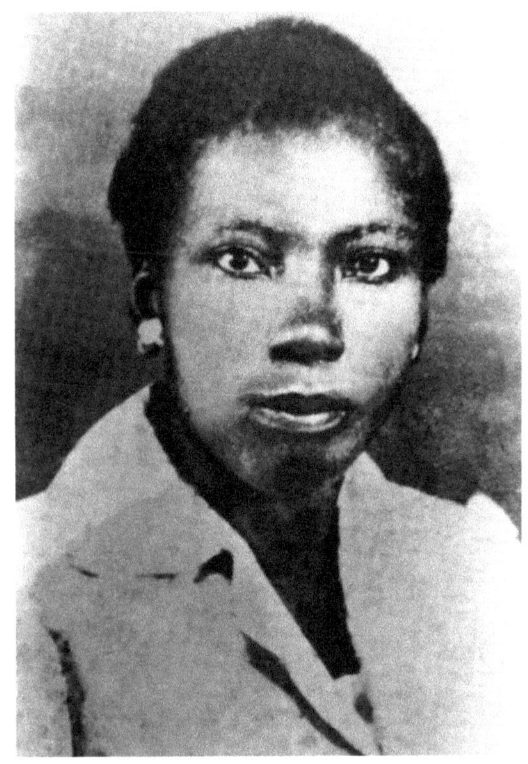
Elma Francois

That same year, confrontation erupted in Rio Claro Village.[24] Here, about three hundred oil and agricultural workers were joined by colleagues from Trinidad Leaseholds Ltd.'s Guayaguayare oilfield, where workers had to live in grass-topped sheds with earthen floors while the manager, Colonel Beaumont, lived in his seaside villa in Mayaro. As the crowd of protestors marched from the railway station toward the town's centre, they were stopped by police. When the workers refused to disband and started throwing stones, the police opened fire, killing five and injuring twenty. More than a dozen people were arrested, among them a woman, Josephine Charles, who was later jailed for a year, alongside her male colleagues.

I doubt she was the only woman there. Probably the most troublesome, though, given that she alone ended up being shackled with the fellas.

Callous disregard for the plight of workers was a feature of colonial rule, and after colonial subjects all over the Caribbean responded with public disturbances, work stoppages, and acts of sabotage which threatened the economic resources of the British economy during the 1930s, colonial rulers were forced to respond. In 1938, the Moyne Commission was appointed by Britain to investigate social and economic conditions throughout the British Caribbean. Their report disclosed deplorable conditions, including issues of child labour and unfair treatment of women who worked long hours for less pay than their male colleagues. Changes were recommended, some of which, paradoxically, were quite adverse to the empowerment of women[25] – something we will touch on in Chapter 5 of this book – however the Commission did recommend a general expansion of franchise. As a result, in 1946, Trinidadian women were given the right to vote on equal terms with men. And yet, it wasn't until 1950 that the Constitution was amended to permit women to assert themselves as candidates for election.

Imagine that: it has only been seventy-three years since the law of the land began regarding us, women, as equal citizens with men.

By the late 1940s–early 1950s, it had become clear that the oil companies of Trinidad could not achieve optimal production on a diet of crude hydrocarbon and man-work. This is why, around each refinery, "camps" were built – stratified according to seniority (which meant "colour and class") – with supportive social *infra*structure to include administrative buildings, domestic accommodation, recreational, medical, and educational facilities. In these camps, women played necessary roles as nurses, teachers, and service providers. These are the very same camps where, decades later, I would visit my friends.

In the oil company offices, however, all clerical and administrative roles had still always been held by men. Only in the 1950s–early 1960s did women begin trickling into the administrative offices as stenographers, as "punch operators" for primitive computers, and as members of the "typing pool". Ellis Lewis recalls the male clerks, still in the majority, making an effort to watch their language around the genteel newcomers. The most senior company men had personal secretaries, Lewis says, and although "you would find a woman in charge of the other women in the typing pool, there was never a case of a woman being put in charge of men."[26]

Shell Apprenticeship Programme

As it became increasingly expensive and inefficient for oil companies to employ a cadre of expatriate staff, they introduced programmes to educate and train local workers. For example, Shell (Trinidad) Limited – the now renamed UBOT – hosted an apprenticeship programme for locals who had reached the age of sixteen. Local boys, that is. Not girls. Only boys.

Prime Minister Dr. Eric Williams speaking to school girls.

But was it so strange not to include girls? This was a time in Trinidad's history when traditional values discouraged the education of girls. Education of sons was understood and accepted, but for daughters, this needed some convincing. The ex-indentured Indian, in particular, thought that education of girls was a wasted investment, since girls were meant to get married and stay at home, not work. Of what value to her parents was an educated girl?[27]

The game-changer on these issues was Dr Eric Williams, who aspired to secure Trinidad and Tobago's political, as well as economic, independence from Britain.[28]

As Premier and then Prime Minister, Dr Williams engineered a new educational pipeline which would, it was hoped, soon begin delivering a flow of competent locals to take control of "the commanding heights of the economy". In his famous Schoolbag Speech, given on the eve of the country gaining Independence on August 31, 1962, he entreated over twenty thousand students during a rally at the Queen's Park Savannah, that the future of the new nation was in their schoolbags. And he was serious: one of his priorities was the steady but gradual democratisation of education, making it compulsory and free to all boys and girls from ages six to twelve, and thereafter on the basis of a competitive examination.

The University of the West Indies.

At the upper end of the pipeline, Dr Williams also supported the establishment of The University of the West Indies (UWI). In 1962, the same year that Trinidad and Jamaica attained Independence, the former University College – an offshoot of the University of London – received independent university status, and the Engineering Faculty taught its debut Academic Year at the Trinidad campus. Trinidad continued to operate as the centre for engineering studies, while Jamaica was the place for geoscience studies. And the UWI, crucially, was equally open to men and to women. It didn't take long for this openness to pay dividends for the energy industry. By the early 1970s, Trinidad had graduated its first female geologist. "In about 1976 or 1977," according to Prof. Andrew Jupiter, the country had graduated its first female engineer. And in 1978, his own wife became "the first female petroleum engineer that graduated with a postgraduate diploma."[29]

But now a new problem arose. Even as Trinidad's educational pipeline was pumping out geologists and engineers, those newly skilled workers had few jobs to flow into. Because, although commercial petroleum operations had been ongoing in the country since 1908, there was, at the time of Independence, no *local* petroleum industry. All the petroleum companies operating in the country were foreign, all the proceeds of the country's natural resources essentially belonged to the Crown and the refineries were managed and staffed mostly by foreign nationals, with locals employed only in junior or menial roles.

In 1963, the government convened a Commission of Enquiry into the Oil Industry of Trinidad and Tobago, later known as The Mostofi Commission. The report of this Commission precipitated the passage, in 1969, of the Petroleum Act and subsequent subsidiary legislation defining "petroleum" to include both oil (liquid petroleum) and natural gas (petroleum in its gaseous form) – and a decision was made to monetise the latter, which was previously considered "waste", thereby expanding many times over, the potential size of the industry (or as Dr Williams called it, "the size of the cake"[30]).

During the 1970s–1980s the foreign domination of Trinidad's oil industry began to abate, and more careers opened up for local workers – both men and women. Commentators have posited that the growing nationalism under Dr Williams, plus labour union pressure from the Oilfield Workers' Trade Union, as well as the Black Power agitation of 1970, were all contributors to the foreigners' departure. Shell (Trinidad) Limited handed over to the government in 1974, resulting in the formation of TRINTOC (the Trinidad and Tobago Oil Company), which continued to expand and evolve as other foreign interests ceded control, and became PETROTRIN in 1993 and then, in 2018, Heritage Petroleum Company Limited – which features prominently in the career of our first interviewee, Ms Arlene Chow (Chapter 2).

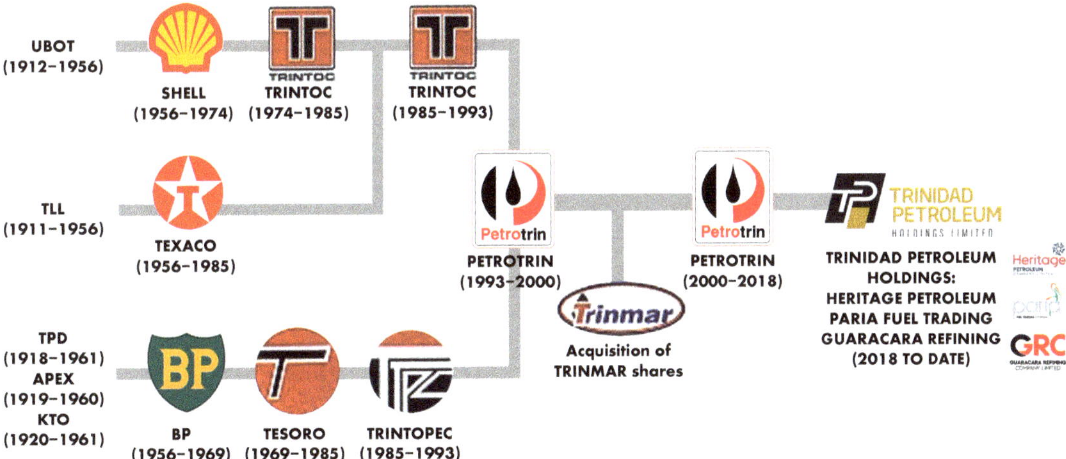

Profits from crude oil were used to kickstart the country's next major industry: natural gas-based petrochemicals, primarily based at Point Lisas in Central Trinidad, which became a major industrial hub in the Caribbean (see Appendix I).

Trinidad and Tobago's oil and gas industry has therefore been in operation for more than a century and now supports the largest hydrocarbon production in the Caribbean, with one of the largest LNG processing plants in the Western Hemisphere. The energy industry is a key pillar for the economy, representing around 45% of its GDP.[31] The upstream sector has seen a return of many international stakeholders, including BP, Shell, Proman, Perenco, and BHP/Woodside.

Women can now be found working at every level in Trinidad's oil and gas industry: from technical and operational, to administrative and corporate roles. They are geoscientists, engineers, legal and marketing professionals, and more. In 2017, women in Trinidad and Tobago's energy sector (petroleum and gas sector, including production, refining, and service Contractors), accounted for almost 20% of that segment in the workforce.[32]

What is most significant, especially when one considers Ellis Lewis' statement that in the old days you would never see a woman in charge of men, is that women now hold senior leadership and management positions in the energy industry. Some of these women can boast of decades-long careers in oil and gas. Currently, both Permanent Secretaries, i.e., the most senior public servants, in the Ministry of Energy and Energy Industries, are women: Ms Penelope Bradshaw-Niles, who holds (among other degrees) a BSc in Chemical Engineering and MSc in Engineering Management from the UWI, and Ms Sandra Fraser, who holds a BSc in Economics from UWI. This year, 2023, also saw the retirement of Arlene Chow, also a UWI graduate, the first female CEO of the national oil company, Heritage Petroleum. And recently, a magazine of the Society of Petroleum Engineers featured Michelle Des Etages, T&T's first female independent oil producers in her role as CEO of Renaissance Energy Limited (REL).

What great achievements! However, I am well aware that neither the single stroke of a constitutional pen nor the seal of a thousand academic degrees can a patriarchal culture change. As my research began to unearth more and more names of women in the oil and gas industry, I became more and more curious about how these women experienced the day-to-day struggle for legitimacy and credibility in such a gendered environment. Oh, what stories they could tell!

And yet, we don't often hear those stories. We don't see them published. What we hear, over and over, are the same stories about the same old male pioneers of the oil and gas industry. It puts me in mind of what the political theorist Cynthia Enloe has said of political and economic history: that it is almost

Early Point Lisas

always written from the viewpoint of men who either actively suppress women, ignore them, or refuse to take them seriously.[33] She writes, "The unquestioned presumptions about what and who deserves to be rewarded with the accolade of 'serious' is one of the pillars of modern patriarchy...[It] offers the chance for masculinity to be privileged and for anything associated with femininity to be ranked as lesser, as inconsequential, as dependent..."

Well, this book is an effort to take women in the Caribbean oil and gas industry seriously, to ascribe to them the agency which they truly possess, and to acknowledge them – whether they work on a rig or behind a desk – not as corollaries, beneficiaries, or victims, but as workers who toil daily and lend strength to the global enterprise of obtaining resources from the earth.

In Section I of this book, I speak with women from Trinidad and Tobago, as it is not only my birthplace and home, but also the most established of the oil and gas economies of the Southern Caribbean. Then, in Sections II and III, I broaden the conversation to include women from Suriname and Guyana, the countries possessing some of the largest oil and gas discoveries in recent years. In conducting these interviews, I sent all the women the same list of questions (see Appendix 2) and invited them to respond in writing, if they felt necessary – one or two did. Mostly, we just spoke, woman to woman, allowing the conversation to evolve organically as I soaked in their wisdom. Some, those no longer actively working in the industry, were more forthcoming with details than others, who still have career prospects to protect. However, certain parallels emerged, which I will discuss in Section IV.

When referring to the oil and gas prospects of Trinidad, Suriname and Guyana, the Trinidadian Minister of Energy and Energy Industries said in a recent interview, "We, as a region, can be stronger as a global player if we're all operating together."[34]

In my opinion, the Minister's words apply equally to the cause of promoting gender parity within the regional oil and gas industry. In this effort, we must *all* work together. Mind you, to highlight the long-neglected stories of women in the industry is to take nothing away from the men. The men have worked hard, and we celebrate them for that. But the women – well, the women bring a different energy. ■

Endnotes

1 Jupiter, Andrew. *Red, White and Black Gold, A 50 Year Journey In The Trinidad and Tobago Energy Industry.* Andrew Jupiter, 2022.

2 Spielmann, Percy E. "Who Discovered The Trinidad Asphalt Lake?" *Science Progress* (1933-) 33, no. 129 (1938): 52–62. http://www.jstor.org/stable/43412353

3 Von Lonski, U.; Syth, A.; Trench, S.; Goydan, P.; Riemer, P.; Fjaeran, T.; Miras, P.; Merchant, W.; Gauthier-Watson, C. 2021. A collaboration between the World Petroleum Council and Boston Consulting Group. "Untapped Reserves 2.0: Driving Gender Balance in Oil and Gas". https://www.bcg.com/publications/2021/gender-diversity-in-oil-gas-industry

4 Scott, J.; Dakin, R.; Heller, K.; Eftimie, A.. 2013. "Extracting Lessons on Gender in the Oil and Gas Sector: A Survey and Analysis of the Gendered Impacts of Onshore Oil and Gas Production in Three Developing Countries". *Extractive Industries for Development;* No. 28. © World Bank, Washington, DC. http://hdl.handle.net/10986/16299 License: CC BY 3.0 IGO.

5 Bridge, G.; Le Billon, P. *Oil.* Cambridge, UK; Malden, MA: Polity, 2017.

6 UN Sustainable Development Goal 5.

7 HomeFront. "Free Yuhself (Give Yuhself Ah Chance)". CD. Bes' of HomeFront: Classics. Caribbean Sound Basin: Sheldon "Shel Shok" Benjamin. 1992.

8 Notable exceptions are: Emerita Professor Bridget Brereton, author of *A History of Modern Trinidad 1783–1962,* Heinemann, 1981; Emerita Professor Rhoda Reddock, author of *Elma Francois: the NWCSA and the Workers Struggle for Change in the Caribbean in the 1930s,* New Beacon Books, 1988.

9 Note 1, *supra.*

10 Lewis, Ellis. *Point Fortin: The Shell Company and a Trinidadian Village.* Ellis Lewis, 2016.

11 Besson, Gérard. "History of the Oil Industry in Trinidad and Tobago". http://caribbeanhistoryarchives.blogspot.com/2023/01/history-of-oil-industry-in-trinidad-and.html

12 Telephone interview with Andrew Jupiter (June 12, 2023).

13 Walter Darwent's Paria Oil company successfully drilled to 160ft at Aripero Estate in 1866 and in 1901 Randolph Rust and John Lee Lum began operations in the same area. They also succeeded, the following year, in drilling to 1015ft at Guayaguayare with eight subsequent successes. In 1906, Arthur Beeby Thompson, an engineer who had worked at British oil companies in Russia, arrived in Trinidad and established a base camp in Point Fortin for the Trinidad Oil Syndicate (later, Trinidad Oilfields Limited). Drilling there commenced in 1907.

14 Note 10, *supra.*

15 Note 12, *supra.*

16 Note 11, *supra.*

17 In 1911, the SS Prudentia departed from Brighton, La Brea with the country's first shipment of petroleum cargo. The oil terminal built by the Trinidad Lake Petroleum Company Limited included a jetty and storage tanks which were the largest in the world at the time. The year 1912 saw the construction of the country's first small crude refinery at Point Fortin, which in 1913, was taken over by United British Oilfields Trinidad Limited (UBOT), a majority-owned subsidiary of the Shell Oil Company. Then, 1917 saw the country's second refinery being constructed at Pointe-à-Pierre by Trinidad Leaseholds Limited (later, Texaco).

18 Note 12, *supra.*

19 "Women in the Oil Patch, Beginnings to 1946". http://www.history.alberta.ca/energyheritage/oil/the-waterton-and-the-turner-valley-eras-1890s-1946/women-in-the-oil-patch-beginnings-to-1946.aspx

20 In-person interview with Ellis Lewis (August 15, 2023).

21 Telephone interview with Grace-Anne Bradshaw (August 16, 2023).

22 Note 11, *supra.*

23 Doyle, Maya. June 16, 2021. "A Lion Amongst Men: The Story of Elma Francois". https://nationaltrust.tt/home/elma-francois-woman-labour-leader/?v=df1f3edb9115; also, see Reddock, Note 8, *supra.*

24 Hosein, Gabrielle. *Trinidad and Tobago Newsday.* June 22, 2022. "Rio Claro and 1937 uprising". https://newsday.co.tt/2022/06/22/rio-claro-and-1937-uprising/

25 French, Joan. 1988. "Colonial Policy Towards Women after the 1938 Uprising: The Case of Jamaica". *Caribbean Quarterly* 34(3/4): 40.

26 Note 20, *supra.*

27 Roopnarine, Lomarsh. 2016. "Interview with Patricia Mohammed: The Status of Indo-Caribbean Women: From Indenture to the Contemporary Period". *Journal of International Women's Studies* 17(3): 4–16. https://vc.bridgew.edu/jiws/vol17/iss3/2

28 Henry, Ralph M. 1997. "Eric Williams and the Reversal of the Unequal Legacy of 'Capitalism and Slavery' ". *Callaloo* 20(4): 829–48. http://www.jstor.org/stable/3299411

29 Note 12, *supra.*

30 Note 28, *supra.*

31 Interview with Hon. Stuart Young featured in *The Energy Year Trinidad & Tobago 2023.* June 15, 2023. https://theenergyyear.com/articles/trinidads-critical-role-in-global-energy-security/

32 Central Statistical Office of Trinidad and Tobago. 2017. https://cso.gov.tt/

33 Enloe, Cynthia. *Seriously!: Investigating Crashes and Crises as If Women Mattered.* 1st ed. University of California Press, 2013. http://www.jstor.org/stable/10.1525/j.ctt7zw2r7

34 Note 31, *supra.*

Appendix 1: Plants in the Midstream & Downstream Gas Sector

Start-up Year	Plant	Product Produced	Plant Capacity (TPY)	Capital Cost (US$ Mn)
1959	Yara (W.R. Grace)	Anhydrous Ammonia	285,000	N/A
1977	TRINGEN I	Anhydrous Ammonia	500,000	1 25
1980	Mittal (ISCOTT)	Direct Reduced Iron	380,000	468.3
1981	PCS I	Anhydrous Ammonia	445,000	333.3
1982	PCS 2	Anhydrous Ammonia	445,000	172.5
1983	PCS Urea	Urea	710,000	173
1984	MI (TTMC I)	Methanol	480,000	183
1988	TRINGEN II	Anhydrous Ammonia	495,000	350
1991	PPGPL	NGL Facility	70 ,000	98.8
1993	M2 (CMC)	Methanol	550,000	200
1994	POWERGEN*	Electricity Generation	1000 MW	N/A
1996	PCS 3	Anhydrous Ammonia	250,000	75
1996	M3	Methanol	580,000	235
1998	PCS 4	Anhydrous Ammonia	650,000	252
1998	PLNL	Anhydrous Ammonia	650,000	300
1998	M IV (M4)	Methanol	580,000	265
1999	Trinity	Electricity Generation	225 MW	150
1999	ALNG I	LNG	3,200,000	930
2000	Methanex - Titan	Methanol	860,000	261
2002	CNC	Anhydrous Ammonia	660,000	300
2002	ALNG 2	LNG	3,400,000	550
2003	ALNG 3	LNG	3,400,000	550
2004	N2000	Anhydrous Ammonia	650,000	315
2004	Atlas	Methanol	1,700,000	400
2005	M5000	Methanol	1,890,000	450
2005	Nulron	Direct Reduced Iron	200,000	180
2005	ALNG4	LNG	5,200,000	1200
2009	AUM I	Ammonia	650,000 Urea - 693,000 Nitric Acid - 495,000	1,200
2010	AUM I (Derivatives)	Urea, Nitric Acid, Ammonium Nitrate, Melamine	Amm. Nitrate- 630,000 Melamine- 60,000	1,700
2020	CGCL	Methanol	1,000,000	1,000

*Formed after divestment of T&TEC assets at Penal, Point Lisas and Port of Spain power stations

Source: National Energy

Appendix 2: Interviewee Questionnaire

A. Intro – General

Please share biodata and explain your most recent Oil and Gas (O&G) job in business terms and in layman terms. Qualifications? Is an engineering degree or STEM background necessary? Is an MBA necessary?

What do you do on an average day? How many people do you supervise? Is the company you're working at a local oil and gas company, or is it part of an international group or conglomerate? E&P? Upstream, downstream? Have you worked in O&G before? How did you get into O&G?

What is your greatest triumph in your career? What would you say is the greatest obstacle or challenge you overcame in your career? Lowest moment?

B. Being a Female Leader in O&G

Did you have female mentors or role models in the industry? Did you have male mentors in the industry? How do you handle the boys' club? Is there a girls' club (should there be?)? Is female competitiveness a problem? What is it like to manage men?

What would you say has been your major challenge working in O&G as a woman? What would be your advice to females entering the industry? What would be your advice to the industry about attracting and retaining females?

Do you perceive a difference (in method or effectiveness) between male and female leaders in the industry? Why are there so few women at the top? Are some sectors of the industry more averse to female promotion/leadership than others?

Diversity programmes, quotas at your job? How effective do you think they are? Are you a risk-taker? Have you ever felt that your worth/competence was questioned simply based on your gender? Have you ever felt lonely and wished there was another woman in the room, on the project, etc.?

C. Impact of Personal Life

Tell me about your childhood. Did you always want to do the type of job you're doing now? Do you feel your childhood/upbringing prepared you in any way for this role? Do you feel that the skill sets you rely on most heavily now were apparent in your younger years? Were there woman influences mentoring you in your years growing up?

Are you proud of your achievements? Is your family proud of your achievements? How has this job impacted your family life? Do you suffer from guilt in relation to your family life? How do you manage that guilt?

Do you think the industry has changed you or your personality? Were you in a different industry, would you be a different person?

NO SMOKING

CHAPTER 2

ARLENE CHOW

Dig Deeper: My Leadership Journey

as told to
Celeste Mohammed

Y NAME IS ARLENE CHOW. I'm sixty-five years old and, just this month, July 2023, I retired after forty years in the oil and gas industry. My mother never wanted me to be a geologist. She said, "That's man wuk," and, in truth, I learnt, early o'clock, that part of the job description for a woman in this industry, is having to constantly prove that you *can* do the job as well as a man.

The Secret of Success – Family Values

Without hesitation, the bedrock of my success has been my family and the values which were instilled in me growing up. I am the last of eight children and we are all accomplished in our own fields: law, accounting, psychology, engineering – even a nun. We were a close-knit bunch – sticking together "like laglee", we were told – who learned very early to share, to cooperate and to consider each other's needs. I remember if you went to eat and there was only one piece of meat left, you had to ask, "Anybody want a piece?" It was even my brother who financially put me through university, and that family support has

Arlene Chow, Retired CEO of Heritage Petroleum

always been with me. I always knew: me and my children would never starve; if I was broke, bored or hungry, my family would be there. That knowledge emboldened me to take more risky decisions in my career.

My mum, Cecelia, was absolutely poor. She grew up in Rio Claro, on a cocoa estate, in the barracks. She was the illegitimate offspring of a mostly-Venezuelan mother and a black father who came from St Vincent and then disappeared just as fast. So, there was poverty and the stain of illegitimacy upon her life, but she was bright in school. She was training to be a teacher but had to stop when she married my dad, as in those days you couldn't teach and be married. My dad's situation was even worse. He had run away from China, at twelve years old, to escape an abusive stepmother. After walking for days, digging up and eating raw potatoes to stay alive, he found his uncle who happened to be leaving for Trinidad, so he hopped on the boat too, and arrived here with neither a cent nor a word of English. He, a Buddhist still going by his Chinese name, Chow-Ka, met my mum, a staunch Roman Catholic, when she was teaching him English. For a while, after they got married, money was so tight that my mum only had one good dress. It was white and she washed it every week to wear again.

My mum really pushed us academically. I think it was because of her childhood, she saw education as the way out, so we never would have to experience her kind of poverty.

When I was five years old, we got our own house in Sangre Grande, and I went to Cunapo Roman Catholic School (to this day, I still say I'm from Grande). We had a "cocoa store" buying and selling cocoa, coffee, nutmegs, copra and tonka beans. I loved going to the store to "help" my father and mother who worked in the business. My father grew most of our vegetables in the garden at home. As a child, I thought everybody got their food that way. At the toughest times in my career, the values that kept me going were those values I learnt at home: study hard, have a strong work ethic and discipline, show respect and make the best of what you have. My mother was big on respect because she grew up being disrespected. She also instilled this: no job is beneath you. She even used to make us go outside and clean the canals in front of the house. Despite all the focus on academics, she taught us that all work is "professional" and that none of us was too precious for hard manual tasks.

My mixed-race, mixed-religious background really shaped who I am and how I approach life. I grew up in the 70s. During the Black Power Revolution, I recall feeling like my family was being attacked from both sides. I was in the

Above: My mother Cecelia Chow née Stephen (nicknamed 'Lily' because of her love of flowers) and father James Chow (Chow-Ka) with four of my older siblings.

Left: My parents.

cocoa store one day when my Chinese father was – yet again! – being harassed by a black man who thought Dad had no right to be a businessman in this country. Enraged, I shouted at the man, "What do you know! My mother is black!", and it was my father who had to stop me from going further. Another time, we went to Pigeon Point, a private beach in Tobago, with a pass that said: James Chow and family. The lady at the gate looked pointedly at my mother, who was in the front seat, and refused us admission. The message was obvious: she didn't belong. Dad had to intervene and say, "She's my wife." Those incidents stayed with me. So, I don't pre-judge people on race or religion at all. In fact, every All Souls, at my parents' graves, me and my siblings say the rosary *and* light joss-sticks to honour our parents and their shared love in the face of so many cultural differences.

School Days in Jamaica

As a student in high school, I'd always loved geography and the earth sciences. So, straight from St Joseph's Convent, Port of Spain, I went to UWI Jamaica to study geology. I was the only woman in the class. I was good at it, I thought I had the respect of all the guys. I genuinely thought they saw me as being as good as them. But when it came to mapping, a highly physical endeavour which involved hiking up mountains and crossing rivers in the Blue Mountains while plotting geological detail, nobody wanted to partner with me.

I realised that my male classmates saw me as a liability in the field. They thought I wasn't physically strong enough. I was really hurt and surprised by that.

To get my work done, I ended up leveraging my academic acumen and my friendship with one of the guys in the class. He didn't have the same love for geology as I did, so we agreed that he would partner with me in the fieldwork, and I would tutor him to pass his written exams. He is still my friend today.

There was one day we had to climb a big ravine. He was a tall guy who could take big steps, so he had already made it to the top of the ravine. I was still in the middle, feeling kind of stuck and trying to make my way up. I remember him looking down at me and saying, "Arlene, I can't help you, you know. Is either you come up or you go down."

Now that I have retired, I can say with certainty, that incident, and his words to me, became emblematic of my entire career in the oil and gas industry: *Is either you come up or you go down.*

In those days – the 80s – the national oil company, TRINTOC (which would later become PETROTRIN and then Heritage Petroleum), would send a recruitment team to the university. My interview proved to be another revelation about what I could expect to face in the industry. When the panel asked if I could drive, I queried whether they had asked anyone else – meaning, any of the guys – and they confirmed they had not. It seemed ludicrous to me because *I* was the one driving my classmates around, as most of them couldn't drive. Yet there I was, the only candidate being asked to reassure the recruiters that driving was in my skill set. Years later, I would remember this and laugh, while gunning a stick-shift Datsun 120Y car, by myself, in the middle of the night, through the muddy backroads of Trinidad's southern oilfields. I swear there were *jumbies* – roaming spirits – in the blackness of the surrounding forests!

Another question of the panel: what would I do if I went on a rig and saw pictures of women in "biological positions" – that's the term the interviewer used. I said I didn't think that would be the appropriate place and that men should confine those pictures to their bedrooms. Most members of the panel seemed to approve of this line of questioning which was aggressive and targeted to remind me that the industry was a man's world which would never accommodate me and my female sensibilities. Afterward, I went home and told my sister, who was doing her master's degree at the time in Jamaica, "I'm sure I didn't get that work." But as it turns out, I did get it.

Early Career at TRINTOC – Geologist

So, I started as a Geologist at TRINTOC when I was twenty-two, stationed at Penal – deep South – where I had never set foot before. I was presented with a contract which said that "men and their wives" would get medical coverage, and I read that thinking it was a mistake, that it was an old contract and hadn't been updated. I was so wrong. "Women and their husbands" were not eligible for coverage. A group of us...maybe five new female hires...drafted a petition and went to the Union, asking for the matter to be tabled on the agenda. However, the Union man's response was, "We have more important things than that to deal with at this time." We tore up our cards, and I never rejoined the Union.

Then, there's the matter of "sexual harassment". Nowadays, we hear a lot about that. But it did not exist when I was a young Geologist starting out in the industry. Sure, it was perpetrated everywhere, but the concept itself

On a boat, heading offshore, well-equipped with charts and snacks.

didn't exist. There was no policy, no procedure for handling it, no guidelines – there was no recourse. So, for me and the other new females, it was hard to settle into our first real jobs. The men, they just see you as "a ting" and they treat you like that. For example, we used radios in those days in our cars and we had call-signs. Whenever one of us would come on the radio and identify ourselves ("This is Charlie-4, going to a rig, etc."), without fail someone - one man or the other - would reply and say the vilest things to us, because it couldn't be identified who was speaking. There was one guy however, a driller, who did defend us. He would get on the radio and say to the perpetrators, "If you are a man, identify yourself." Not surprisingly, no one ever did. I have never forgotten that one brave ally. Just like I've never forgotten the generous men who shared their Sunday *callaloo* with me on the rigs, or who would come out and find me whenever I got lost driving through the spooky oilfield roads and ended up instead in a cow-field. "Arlene could never find a rig!" they used to say. There were lots of good men.

But there were other incidents of harassment: I have had my hands held way too long; I have been kissed on the neck. I have even been advised, by someone who was well-meaning, that since my boss was "sweet on me", I should capitalise on that, and "do the things that men and women do in the night and forget the next day." And, because I was twenty-two, fighting to get my bearings in a first job and not confident in myself, I didn't know how to tell those men, "Haul your tail and leave me alone!"

It was very lonely. These men had their own clique: they'd drink rum together in the rum shop, they'd lime and party together. And we, the few women, had to fight through this alone because the very thing that was causing us all this stress was like a bogeyman that no one else could see or would even address.

There was a distinct hierarchy and we felt like we were on the bottom. There was one toilet for senior staff and one shared by everyone else. I vowed to myself that if I ever got the chance to lead, democracy would extend even to toilets.

Aboard ship with my Executive Assistant, Natalia Jacelon

I got married at twenty-three, got pregnant, lost my first baby and then I became pregnant a second time. I informed my boss and asked for some consideration in respect of assignments, thinking that if I could have fewer physical stressors in the early stage, I would have a better chance of taking the pregnancy full-term. Instead, my boss said, "Unless I see a medical, you have to keep going on the rig." In those days, all they had on the rig were these porta-potties and they were horrible. When you're pregnant you need to go to the bathroom so often, I would really have liked some consideration, so that I could avoid all that. But I had to keep my job and do what my boss ordered. Eventually, I did get a medical saying I was pregnant. He reassigned me then, but immediately cut my salary by 15% because I was no longer going on rig duty.

Eventually, I had my two kids while still at the national oil company. As my life evolved, I also saw the company evolve into PETROTRIN and I was moved from Penal to Pointe-à-Pierre. I also ended up with a different manager who was much more understanding. Still, it was hard balancing my job and my home life. We were living in St Augustine, and I had to reach to work for 7:00am. I never did really. I was able to strike a deal with my manager that I would come for 8:00am and work through lunch. I tried putting my kids in daycare but that wasn't working out, so I had to get a nanny to look after them. I spent many panic-stricken mornings, pacing in front of the house, hoping she would arrive on time or even reach at all, so that I could get to work on time.

Further Education in Florida – Master's Degree

After twelve years in that Geologist role, I left to pursue a master's degree in engineering at University of Florida. My children were still young, so they had to come with me. We sold our apartment in Trinidad to fund the trip, but things were still rough and both money and time were tight. My spouse at the time was also studying, and I was working as a teaching assistant to help offset tuition. My daughter was of school age, but my son was still too young, so I had to put him in a daycare. I couldn't afford that every day, so I would send him there for three days a week, then I'd stay home on the other days. I had to develop coping mechanisms, making sure they were asleep by 8:00pm then using the rest of the night to study. I wouldn't say we were living hand-to-mouth, but every day I would wear the same old jeans and sweaters, with my hair in a messy ponytail, as it cost too much to go to the hairdresser. So much so that the one occasion I did "dress up" – bought a dress, cut my hair, put on makeup – to go defend my thesis, my professor did not recognise me, and passed me straight on the corridor. It was during that time I developed a bad habit that persists to this day: Coke and Cheetos. By the time I got those kids up, bathed, fed and off to school and daycare, then got to the Engineering Department, the only thing available for me to eat was whatever was in the vending machine. So, I subsisted on a diet of Coke and Cheetos. I call it the breakfast of champions.

I was the only woman and the only black person in the master's programme. This was during the First Gulf War and, because of the frame of reference most people had, I was often mistaken for an Arab. Precisely because of how singular my presence was, my teaching assistant job was given as part of the school's diversity program. When I graduated, the Diversity Dean was adamant that he wanted me to "walk" at graduation. I guess it would have made a very public statement and triumph, for both the school and for me. However, I couldn't afford it, because by then I'd returned to Trinidad, and it would have meant incurring all the costs associated with flying back up to Florida.

The Return to PETROTRIN – Information Systems

When we returned home in 1994, I had a Master's Degree in Engineering and a Diploma in Information Management...but nowhere to live. So, we stayed with my sister until we were able to save enough to buy the home where I still reside.

I went back to PETROTRIN but switched from Geology to Mapping and Information Management. The new role called on all aspects of my training, as a geologist with experience in mapping, as an engineer, and as someone familiar with programming. It wasn't a senior position, it was four of us in a room – a tight-knit team – just putting our heads together and getting the work done, i.e., developing and running PETROTRIN's entire Well Information System. I enjoyed those days – they were probably my best years at PETROTRIN – and I've kept in touch with the team, from back then. Imagine: the system we built still exists today.

The Move to BP – First Leadership Positions

Then, I was headhunted by BP, which was starting up its computer and information system for the exploration department. I moved across in 1998 and was eventually promoted to Head of Exploration and Production Computer Information Systems, and that was my first leadership position. I was happy doing that. However, the company needed someone to lead their Subsurface team in drilling, reservoir engineering and geology. They asked me to do it. I said yes and that started a pattern where I found myself changing roles almost every two years while at BP. I think that's because I never back down from a challenge. I always say yes. My first boss in BP told me, "Proceed until apprehended," and that has been a mantra throughout my career.

At some point, I became interested in the Delivery Manager role at BP, which would involve responsibility for running all the oil fields. I knew, though, that to make myself the obvious choice for the role, I had to learn Operations. The challenge was how to do so without jeopardising my substantive Monday-Friday job as Subsurface Leader. So, every Saturday, I would use my free time to go offshore and just immerse myself in Operations, learning that new side of the business. It paid off, because when the Delivery Manager job became available, I was selected. That's what kickstarted my Operations career.

I became Delivery Manager and performed well in the role because I have always been a lifelong learner. It's never been surprising for me to see that trait displayed in psychometric tests throughout my career: 100% lifelong learner. I'm even thinking about what to do now that I'm retired. Maybe I'll learn some AI (artificial intelligence) next!

At BP as Delivery Manager.

People tend to say that I'm a good speaker. However, I am an introvert; public speaking didn't come naturally to me. It was, again, the result of my eagerness to embrace new learning. The company sent me for media training: it's just you, a man with a camera, and the coach. They taught me how to use my voice, how to ad lib, and most importantly, how to tell a story. I've fully embraced the idea that storytelling is crucial to successfully conveying a message, and I've seen it work throughout my career.

As Delivery Manager I had some tough times as well. There were things that kept me up at night, e.g., handling a divestment where we sold the entire oil asset of BP and I had to send home eighty-two people in one go. This affected me deeply as these were my guys and I loved them. It was probably one of the hardest things I'd had to do in my career, up to that point. Who knew harder was coming!

I was promoted to Vice President of Corporate Operations in bpTT responsible for the support functions: Health & Safety, IT, Facilities and Performance Management. This was my first taste of executive leadership, where I sat to make decisions with the President and all of the VPs in bpTT, a team of strong personalities, each wanting to have their own way. It was a difficult role for me. I had to adapt from being a command-and-control leader to being an influencer of other leaders. It's hard convincing people to do what *you* think is best.

In that first VP role, while learning about the strategy of the company, I had to learn how to execute that strategy, even at times when I did not entirely agree with a course of action. bpTT was, at that time, trying to improve its safety culture and, as the VP with responsibility for safety, I often found myself in contentious situations. One time, discussions got so heated between me and another VP that it almost became a brawl refereed by the Regional President, who ended up telling us, "Take it off the table". As we left the room, I turned to my colleague and said, "If you want to fight, bring it on. But if you want to work together, we could do that too." After a while, we developed an amazing working relationship.

It was during that VP Corporate Operations role that I wrote my first pandemic plan because the company was worried about the Bird Flu at the time. That was an immensely frightening task, as I came to see that the country was not ready for any kind of pandemic. The experience served me well, though, many years later during the COVID-19 pandemic.

Senior Management – The Alaska Years (2009–2011)

After VP Corporate Operations, I was in line for VP Operations – responsible for all offshore fields and production – which would have represented a major promotion from local level to Group level leadership. But that didn't happen. I was told outright that I didn't get the job because I was "too tough" but that I should take that as a compliment. I was confused, because I was a black woman in a male-dominated Anglo-American company, trying to lead in a role which has traditionally been held by men. What was I supposed to be, if not tough? I still haven't figured it out!

I was offered an international posting instead: Alaska or Azerbaijan, to prove myself, I was told. I applied for both and got the job in Alaska. I remember laughing and thinking: *how ironic, maybe there was need for a tough woman in the Arctic.* At the time, it was deflating not to get the promotion, but Alaska turned out to be one of the best roles in my career. Alaska really is the last frontier. It was a new life for me in many ways, living and working 250 miles north of the Arctic circle. We called it the 'slope'. Imagine: my daily commute started by plane, then I took either a helicopter, a boat, an armoured vehicle or a hovercraft, depending on the season and the condition of the ice. I lived on a camp in Deadhorse (yes, that was the name of the town) for four days every week. Only on weekends did I go home to Anchorage.

During my years in BP Alaska, I had three different jobs. I went in as Infrastructure Manager, and almost immediately faced my first challenge. There was an earthquake and a volcanic eruption, the snow turned black, and it became completely dark, with temperatures dropping to -60°F with wind chill. I had barely unpacked and I had three thousand very unhappy guys stuck on the slope as the planes couldn't fly. It was, as we say in the industry, like "drinking from a firehose" to try to house, feed men and gradually get them home. As if that wasn't enough of a baptism, worse occurred. During a period when I was holding on for my boss, who was on vacation, there was a fatality. I had to take control of the emergency. Some of my decisions, particularly as I was a newcomer, weren't well received. The men on the slope had gone through so much trauma over the years that they couldn't function in the immediate aftermath of the fatal incident, and I was insisting that they conduct the necessary incident management and shut down activities. The slope is an unforgiving place. They thought I was being heartless when I said, "Our colleague has died. God rest his soul. Now we need to protect the living. We will grieve later." I remember the resistance, the bad-mouthing, but I had to do what was necessary to ensure the safety of the other workers in my charge.

And I did grieve. The psychologist who came to the slope after the fatality called me "Mother Earth", as I had to be tough and empathetic at the same time. We had a funeral on the slope: the day was a total whiteout with snow and the temperature was about -20°F. I looked out my office window as the men and vehicles – about fifty trucks and snow ploughs included – drove with lights and horns blazing, the American and Alaskan flags flying, as they passed the area where our colleague died. I was wracked with sobs. I felt a pain, all the way to my stomach, for the life of a colleague and the weight of the lives of three thousand men. Later, the same men who had resisted me in the immediate aftermath, came and apologised, acknowledging I had done the right thing.

As a result of how I handled that incident, the Regional President gained confidence in me. I was given the job of reorganising the whole of the Alaska operation into a functional model. For six months, I was holed up with a small team, redesigning the organisation. Although the ultimate decision on what the company would look like was not mine – all I had was some influence on the redesign – my role rendered me a kind of pariah. When people passed me on the corridors, they wouldn't look me in the eye. They knew that redesign meant some people would have to go home. It was a thankless job but I'm proud of what I accomplished with my team, as we treated people with respect and dignity. Subsequently, I was given the job of Area Operations Manager for all the North fields in Alaska.

It may seem like the twenty-two-year-old Arlene, who didn't know how to respond to sexual harassers, had disappeared, transforming instead into a confident, tough woman. But, if I'm honest, I have to say I was scared when I arrived in Alaska. In the middle of my heart, I was terrified. I was a woman – a black woman – directly in charge of my team of three hundred white men! I was truthful with them; told them I knew I had a lot to learn from them and I was counting on them to teach me. I was humble, and they responded to that, once they realised I was sincere. Humility is necessary to make it through these kinds of challenges. In time, they admitted that they had looked me up on the internet and were as scared about me coming as I was about meeting them. They thought, "She had to have done something really wrong in Trinidad, to be sent to Alaska!" In the end, we were able to laugh it all off, and I was happy and proud of the work I did over there. I loved Alaska. I felt I had done something right and, when I returned to visit in the later role of Chief of Staff, I was greeted in a way that made me forget the former days of unbearable cold. "Welcome back, Arlene," they said, and that warmed my heart.

Working at North Star in Alaska, a man-made island.

Chief of Staff – The London Years (2012–2014)

At the close of the Alaska assignment, I was, at least to my understanding, due to come back to Trinidad to assume that elusive VP Operations role. However, yet again, it was taken off the table, for reasons unconnected with my performance. I was blindsided. I felt letdown and betrayed.

Thankfully, I was able to leverage my past performance, my knowledge and experience, and the network I had fostered within the wider ranks of the company's global leadership. I was offered a post at the Corporate Office in London, as a Chief of Staff to the Executive Vice President who was responsible

With my sister Annette and my niece Aisha in front of the Wall of Excellence of St. Joseph's Convent. We all three have our names on it.

for managing the entire production of the global company. My marching instructions and job description from my new boss were, "You will figure it out." So, I had to: from strategy writing, speeches, presentations, liaising with the other Chiefs and Regional Presidents. And, because it was a global role, I travelled extensively and had to be available almost 24/7: apart from London office hours, somewhere in the world was always now waking up and needing something. The only other piece of advice that boss ever gave me was, "In this role, the train always arrives." I took that as a guiding principle for not only that job, but every role after it. I've tried to create an organisational culture where we deliver what we promise.

I'm not sure I could have done that London job if my children had still been young. By then they were grown and at university and would come visit from time to time. But then again, maybe I could have done it, even if the children were young. I certainly would have tried. That's how I am: willing to try anything. Sometimes, I even feel like people hold me back. I once wanted to jump off a cliff in Ricks Café in Negril – a recreational jump, of course - but my kids intervened. Another time, while whitewater rafting in Alaska, I wanted to jump off and feel the freezing water, and my family intervened again on that occasion. Maybe that's why my psychometric tests also say I'm 100% "eccentric" – I just want to explore, even if no one else comes along.

As COO of ALNG

Speaking of family, my career really put pressure on my first marriage. My job monopolised my time and took me away from the family. We married very young and then found ourselves diverging along different paths. This presented a dilemma that cut to the core of my life. I *like* to work, I was passionate about the type of work I was doing, but me being happy at work meant making my partner unhappy at home. And to make him happy, I would've had to dishonour my growing sense of who I am. We got divorced. I ended up re-marrying while in Alaska.

Home Again – Chief Operating Officer, Atlantic LNG

After the Chief of Staff role in London, I came back to Trinidad and was assigned to Atlantic LNG as Chief Operating Officer (COO) – the first female ever, in that role. I remember on my first day, they were in the middle of a turnaround – where you shut down the entire plant and conduct maintenance, etc. Before introducing myself to the workers on site, I decided to consult the Head of Safety as to whether there was any particular message he wanted me to convey. Should I be technical or inspirational? He said, "I want you to be a bitch." After my initial shock, I burst out laughing but I understood immediately that he wanted me to establish myself as a serious woman whose

Yearly kick-off Workshop at Atlantic LNG

stipulations should be followed, so we could complete the turnaround without any safety incidents. I replied, "Leave it to me. No practice necessary."

You would think that this would be an easy role for me – perhaps, operationally, it was. However, in terms of organisational politics, it was one of my hardest jobs. Atlantic LNG was a joint venture with leadership from all the major oil and gas companies in TnT, and I was a secondee from BP. We all came with different, sometimes contrasting agendas and cultures, and the difficulty for me as COO was aligning and melding those into a workable model.

That was one of the roles where I had to deal with sexual harassment again, but this time from the perspective of a senior leader receiving and handling the claims. There were cases of physical groping, there were also cases of stupid pranks, e.g., sticking obscene signs unto a female's back. Things like that always had a triggering effect on me, because I had gone through it myself as a young woman in the industry. Sometimes, my response as a "boss" receiving such complaints, was to admonish the male staff in language that the lawyers were unhappy with ("I will send you home and pay for your taxi too!") but I felt it was necessary to give a clear signal that sexual harassment would not be tolerated on my watch. I thought it was ridiculous that in the 21st century, men were still behaving that way.

The Second Coming, Heritage Petroleum (formerly PETROTRIN) 2019–2023

With Prime Minister Dr. Keith Rowley and Minister of Energy and Energy Industries the late Franklin Khan at the at Trinidad and Tobago Energy Conference and Trade Show, 2021.

In 2019, I had already retired and was in Spain with my present husband when I received a call asking me to return to Trinidad, to hold over for six months, as CEO of Heritage Petroleum. After some discussion with my family, I decided to do it – to give back to the company that had started my career and trained me in the earliest years – because it was supposed to be a short stint. Unfortunately, COVID happened, so my tenure turned out to be a much longer one.

It was tough leading in COVID time, with all the uncertainty and misinformation. Unfortunately - or fortunately - my experience with pandemic planning came in handy and I was able to share that knowledge with other companies in the group. On many days, I was making life-and-death decisions. We had to keep the company running but preserve the health of the teams in the field that were risking their lives every day. I remember having to repair a hospital in a month and get it ready to receive patients - we did it! But what really touched me was the support, freely given, by all the nation's oil and gas companies and contractors. If only we could always work together like that, what would our country look like?

Now, in 2023, as I have ended my career (for the second time) I see that leading Heritage was an opportunity to close other loops that had started back when I was a twenty-two-year-old graduate Geologist. This time, I went in with two objectives: to make sure the train always arrived, and to dismantle the hierarchical culture that had once made me unsure if I could speak up against harassment. I never forgot the toilets for "us" and "them".

I had a great team. We put together new systems and procedures and started from scratch. It was like building and flying the plane at the same time. I

Guayaguayare Team with their garden from which they donate to the needy.

"flattened" the organisation. "Don't call me anything but 'Arlene'," I told my teams, because I wanted every person to know that the way forward was about collaboration and that their roles and voices were important. I saw it as crucial to change the old PETROTRIN culture of "superiors" and "inferiors" – which had probably been ingrained since the earliest days of the oil industry in Trinidad, when everything was foreign-owned and stratified according to colour and race, local and expat. But I believe that dignity, respect and appreciation are basic human needs that everyone deserves.

People remember how you make them feel. There was a day at Heritage when we had an unfortunate tank failure. The team that worked the incident told me, long afterward, that they'd felt empowered and motivated to get things done because I had shown up and said to them, "Do what you have to do. I have your back."

Heritage has done well; I think the most recent figures show we contributed 4.2% of Trinidad and Tobago's GDP. We made billions for the country, but when I was driving out for the last time, I was startled to realise that the staff had all along been judging me by other criteria. One of the security guards at the gate hugged me and said, "Thanks for all you did for the country. One thing for sure: nobody could ever say you take anything. You gave."

Heritage was the job of my lifetime, and I am grateful to the Chairman and Board of Directors who gave me the opportunity.

Opening the Heritage Safety Village with Heritage Directors and Leadership team, 2022.

Leading and Managing in a Gendered Environment

I don't think, as a manager, there's any difference really between managing men and managing women. At the heart of it, I see my role as setting expectations and giving people direction, support and empathy. I see myself as an easy-going boss, but the employee – man or woman – has to deliver. I have never really had very many women reporting to me because I've typically worked on the technical and operations side. However, regardless of gender, I just want to get the job done, and, like my mother, I do not tolerate disrespect.

In hindsight, I was angry – and stayed angry – for a while because of the disrespect I had endured, constantly fighting up as a young woman in the industry. And I think my anger was prolonged because, in some ways, the offensive language never went away, the words just changed and got more nuanced as I advanced through the leadership ranks. It's not an exaggeration to say that, in the course of my career, I have been called every adjective known to womankind: not hard-ass enough, too tough, too emotional, not calm enough. Qualities which, in a man, might be described as "passionate", "decisive", or "driven". Honestly, I've often wondered what it is they really want from a woman leader within this industry. I wish I could say something different – it, really pains me to say this – but even for me, after all my achievements, the oil and gas industry remains a tight-knit boys' club.

I have never had a female boss. There is such a dearth of senior female leadership in the Operations side of the industry that, when I became a Chief of Staff, women from around the world asked me to be their mentor. It was a lot for me to handle. I always tried to make myself available to the women back home, in Trinidad and Tobago, so they knew they could consult me informally at any time. To add another formal layer of mentorship with so many other women across the globe, was going to be difficult. I remember dropping my head on the desk and saying to my boss, "I have no more advice to give. Can't you see there is a need?" In the end, I could only have accepted two formal mentees, and I felt bad refusing everyone else because I understood the attraction: I was the only black woman at that time to hold a Chief of Staff role in the "upstream sector" and they wanted to know what I had done to reach where I did.

The Challenges of Being a Woman in Oil & Gas

My major challenge working in oil and gas as a woman was getting people to respect me and to believe that I knew as much as them, and that I could do the job as well as them – or better. It wasn't easy. I was fortunate to have three male mentors who held senior leadership positions and who believed in me and supported my career.

However, there have been times when I've entered a room with only men – many times, only white men – and found that nobody wanted to hear what I had to say. I'm not suggesting that the challenge was confined to white men. The challenge was the very fixed and limiting perceptions – held by both men and women – of what a woman's role, function and disposition should be. That barrier was always there, in every room, and I had to break through it to be heard. I admit there've been times when I've been a coward and felt that maybe I should shut my mouth and just bat in my crease instead of speaking up. But eventually, I had to find a way to assert myself, in order to do my job effectively. Proving myself over and over again became part of the daily grind. For example, there was a time when I would go to meetings accompanied by one of my direct reports, who happened to be a white male, and consistently people would approach *him* on the assumption that he was the boss, and I was his assistant. It became a funny game we would play every time: going along as people chattered to him as "boss", and then springing the surprise on them.

In the last few decades, things have been improving to facilitate women's entry into oil and gas. However, I do remember seeing a statistic that even when there'd been something like 40% female hires within the organisation, there was only a ridiculous number like less than 10% at the top decision-making levels. Those numbers suggest there's been a culling of women who attempt to ascend the ranks of the industry. I think that's right: there is an invisible sickle. It's the constant psychological pressure to prove yourself at work (*swipe!*), it's the fact that when we get married and have kids, the responsibilities of spousehood and parenthood tend to fall heavier upon us than upon the men (*swipe!*).

Women need support, they need coaching, they need other people to talk to about the perils of being a professional woman in a male-dominated environment. We women also need to forgive ourselves for not being perfect, for not meeting the standards imposed on us as to what a woman should be. I did not do everything right as a mother; there's a lot of guilt there. A failed marriage – that's not something to be proud of either. And when I think of every big mistake I've ever made in my career, I cringe because each can be traced back to me forgetting to be humble. But I have been working on forgiving myself.

Regarding diversity programmes, I am a bit ambivalent. I never want to be given a job just because I'm a woman, but rather because I'm the best person for the job. The thing about corporate programmes and policies is that they can be very successful in dealing with overt harassment or discrimination of women, but I'm not convinced of their effectiveness in dealing with micro-aggressions and the more covert forms of harassment and discrimination. So, the question is: how to change not only behaviour but entrenched ways of thinking about and reacting to women? Can a corporate programme or policy accomplish that? I'm not sure. Also, in this part of the world, women face all kinds of social and cultural biases, some of which aren't openly acknowledged. As if being a woman isn't hard enough, there's the issue of whether you're a light-skinned or dark-skinned woman. There are people who will brand you as "front office" or "back office" depending on your skin tone. I've seen this happen. So, one of the things I've had to do, on a personal level, to address the issues I've perceived, is step in and act as a promoter of other women in the oil and gas space, i.e., to defend their capability and their brilliance, and to add my voice in support of theirs.

Above: With my family. Left to right: My daughter Akela, my husband Thor, my son Mikkel and daughter-in-law Britney.

The loves of my life, Camden and Avery, my grandkids.

At Ascot with Thor.

Ultimately, to any woman coming into the oil and gas industry, I would recommend the words of Langston Hughes:

Hold fast to dreams
For if dreams die
Life is a broken-winged bird
That cannot fly.
Hold fast to dreams
For when dreams go
Life is a barren field
Frozen with snow.

Hold fast to your dreams and work them. Lean on your values, forgive yourself and others. I am forever grateful to my family, my children Akela and Mikkel, my husband Thor, and my sisters and brothers, for forgiving, supporting, and loving me on this crazy journey.

I try not to hold on to bitterness because that brings me down. I have faced rejection like anybody else, and in the immediate aftermath of it, I have spiralled into bouts of hurt, anger, and self-doubt, but ultimately, I pick myself up and make the best of what's in front of me, and I try to let things go as quickly as possible. I don't hold anybody in mind. My mother had to pick herself up and find a way, my father had to pick himself up and find a way. When I think of what they endured, my struggles seem small in comparison. I just tell myself, *dig deeper, Arlene.* ■

CHAPTER 3

GISELLE THOMPSON

It's Not A Small World After All

WHEN A FEMALE VP of a multinational energy giant talks about her career and says, "Funny story...", it's unlikely to be "funny" in the ha-ha sense, but rather a story that discloses some irony of womanhood, some strange or odd obstacle she's had to surmount, by sheer force of will, along her trek into the rarefied climes of senior leadership. That's what I'm bargaining for when Giselle Thompson, VP Corporate Operations at BP, the London-based oil-and-gas "supermajor", utters the narrative preface. Hardly ever, to my recollection, have such words been followed by the tale of a woman with a deep concern for vulnerability.

Yet, that's the story Giselle shares, during a video call of such intensity and duration that it exceeded my free minutes and had to be continued via WhatsApp, as we juggled our Friday afternoon schedules. Having won such a solid block of Giselle's precious time, whatever else I had on my agenda would have to yield. It was crucial, for several reasons, to get her perspective. While Chapter 2 of this book focused on a woman-leader on the technical/operations side of Trinidad's oil and gas industry, Giselle is representative of the global phenomenon where most of the women in oil and gas are concentrated in business and administrative roles. Which, of course, begs the question as to how vastly the experiences of suits-and-heels women differ from the experiences of steel-toe-boots women. Further, if you think of BP's operations in Trinidad

Giselle Thompson

as a pie, half of that pie involves technical risk, the other half is non-technical risk. Giselle, as VP of Corporate Operations, is responsible for managing much of the non-technical side. Hers is a massive job.

How did she get there? And was this pretty, petite woman the type who talks softly but wields a big stick? I simply had to know.

Previous Career – Brand Manager

She describes a career which begins as Brand Manager at a major pharmaceutical company. Lots of travel, creativity, and excitement. Fast-moving consumer goods were her thing. She was young and ambitious, she set herself a personal goal of obtaining her MBA, she paid for it with her own money, and she started the course.

"But it took me twenty years to finish," she says. "Yes, I started, and then I asked myself, 'Do I really want to spend the first years of marriage with my head in a book?' "

Giselle shelved the master's programme, with a promise to revert soonest. "Then the penny dropped," she says. With the coming of her first child, the exigencies of family life took over, and almost overnight, the travel and excitement of her job began to cost her those one-off moments – the first smile, the first tooth, etc. – which can never be reclaimed.

Then, an opportunity at BP was floated. Back then, like many people, Giselle didn't have an appreciation for the breadth of roles, jobs, and careers available in energy. Instead, she'd always assumed that working in oil and gas would require a science, technology, engineering, or mathematics (STEM) degree or qualification – none of which she possessed. "Where can I fit into this world of energy?" she asked herself.

Transition into BP – More Branding

However, much like her, BP was, in the early to mid-2000s, reinventing itself. Although the company had operated in Trinidad and Tobago for a long time, and had done much to build its brand, it was still evolving out of a merger with Amoco in 2000. "They wanted someone with a marketing background to help envision and enhance the position of the brand in T&T, as a local energy company rather than a foreign multinational," she explains. On a most auspicious day – her daughter's first birthday – in 2005, Giselle made a lateral transition into the company, as Brand Manager.

She might not have had any experience in oil and gas, but what she did have by the barrel-full was ambition and drive. She also came with a two-pronged personal credo: (1) stay curious, i.e., get to know and understand everything you can about the business, about the people involved, then build those relationships; and (2) add value wherever you go. This, Giselle insists, is what helped her scale the steep learning curve she encountered at BP. "It was so daunting at first. It was a very different industry to consumer goods. It's a very technical industry and the first challenge was to understand how it works."

She soon found her footing, became comfortably ensconced in her role, and was pregnant with her second child when the position of VP Communications and External Affairs became available. When the incumbent asked Giselle if she intended to apply, she replied, "Are you sure I can do the job? I don't think I have all the skills. I came here to do brand and marketing. I don't know anything about corporate responsibility, external affairs, government lobbying, and all that stuff."

Towards the end of her maternity leave, she was invited to a meeting by the then CEO, to which her response – so confident was she that she was not in line for anything other than a perfunctory welcome-back-to-work debriefing – was, "I coming back just now. I'll be back there in about a week or two. Can we catch up then?" However, at the CEO's insistence, she went to the meeting and was offered the VP post.

Senior Management – Becoming a VP

She had an explosion of self-doubt. "I actually started to talk myself out of the job. Because I had a two-year-old and a two-month-old and now this was a huge role. I wasn't sure I was ready for it or that I could manage both a very young family and the heavy responsibilities of such a senior role."

One could say that Giselle's reaction was typical of anyone, in any company, who is called to step into a larger role. But this is BP, the world's third largest supermajor, and the principal gas producer in Trinidad and Tobago, a country dependent on gas for the majority of its revenue. This is BP, the company which has enjoyed the longest association with Trinidad and Tobago of any energy company operating in the country. Except for a gap between 1979 and 1998,

With an employee from BPTT's Mayaro-based MIPED microfinance programme

the relationship spans 71 of our nation's first 100 years of commercial petroleum production. Giselle was being called to manage exactly that: a long-established, multi-party relationship involving not just the government, but also regulatory agencies, a whole town and geographical district (Guayaguayare/ Mayaro), and, I daresay, the health and economic well-being of every citizen. No wonder she was scared.

She tried selling her boss another solution: splitting the role between her and someone else. But he refused, saying, "You can do this role, you will learn, and we'll give you support."

It was his patience as she worked through her misgivings, his articulation of reasonable corporate expectations – it wasn't a sink or swim situation; he knew she'd have to grow into the role – and his pledge of company support, which convinced Giselle to consider the opportunity. After discussing it with her husband and parents and receiving their emphatic, "Just go for it," she accepted. However, there was still another area of trepidation to overcome, as she'd been offered the VP role over a male colleague who was senior to her, both in age and tenure. She had concerns about how he would accept and relate to her, since overnight she'd gone from "colleague" to "boss".

"I decided to be honest and vulnerable. I let him know that I respected his experience and understood that I would need to learn from him. I let him know I wasn't coming to tell him how to do his job, but rather I would be leaning on him to help me do mine. And we developed a good relationship."

The first year as VP was still extremely difficult, and Giselle lived with a constant fear of disappointing the team, the organisation, and the people who'd believed in her. There was much to learn on the job while, at home, she was raising a newborn and a toddler.

As I listen to this part of Giselle's story, I wonder if she realises that what she's describing is a quintessential career-derailing season of the kind identified by McKinsey & Company.[1] According to their research, women begin to drop out of the oil and gas industry at two key inflection points: (1) that first promotion from entry level to lower management; and then, (2) the leap into senior management.

We discuss this research for a bit, and she explains, "Yes, it just seems like at exactly the same time as you're growing your family, your career begins to grow. When you're growing your career, you really want to double down. You want to spend more time learning and understanding. You want to go all in professionally, but at the same time, you have these huge family demands. You have young kids. And sometimes it's not just kids, it may be older parents to care for. Many women aren't able to find a balance, so they make that hard decision to put family over career."

"Do you know any women like that? Who opted to transition out of the industry?" I ask.

"I do. I know of a few women. I think, for them, the family sacrifice may have been too much. And I totally get that because believe me, I've had many moments when I've said, 'Why am I doing this?' But then I keep on doing it, I keep pushing forward. But it's a real challenge."

"So why do you think you survived that career near-death experience?" I enquire.

Support Systems and Role Models

Giselle's reply: it really does "take a village". She gained perspective and invaluable advice by chatting with other females who'd been there before, in the business world. She is a member of the International Women's Forum, a global professional women's organisation focused on female leadership and developing the next generation of female leaders. The Trinidad Chapter was started only a few years ago but has provided Giselle with a support system of senior female leaders from whom she can learn – and trade war stories. They helped her realise that the tough times don't last forever. "So, it's about sticking things out. In a year or two, you'll have a better grasp on the job. You'll know where to focus and where not to focus. And then the kids grow – in each stage there are new challenges, but it does get easier."

Giselle also admits she might not have made it without her family support system: spouse and parents, extended family and friends who did their fair share of pick up and drop off.

With her parents and sisters.

Here's where our conversation digresses into musings about the power of precedent and role modelling. In short, role modelling signals what's appropriate and what's possible. According to Giselle, she grew up in "an average middle-class family" where her mother was a career banker, who spent her entire life working while also raising four daughters. Her mum was not at every practice and game, but there was a family system that made Giselle and her sisters still feel supported. "I saw that my parents shared parental responsibilities. My grandfather was also there with us, every evening, helping with homework. I saw that you can have a professional career and you can have a family, you don't need to choose."

Giselle's assessment of her parents' impact is precisely in tune with the findings of a recent Harvard Business School study[2] across twenty-four countries, which proves there are very few things that have such a clear effect on gender inequality as being raised by a working mother. Women whose mums worked outside the home are more likely to have jobs themselves, are more likely to hold supervisory responsibility at those jobs and earn higher wages than women whose mothers stayed home full-time. Men raised by working mothers are more likely to contribute to household chores and spend more time caring for family members, because growing up, what was being modelled for these sons was the idea that you share the work at home.

Giselle also describes an atypical parenting style in her childhood home – not your usual West Indian authoritarian[3] regime – and credits this, in part, for her success and resourcefulness. "They weren't helicopter parents or strict disciplinarians who tried to control everything. They tried their best to guide and coach and invest and cheerlead, but they never pushed or demanded what we should be. They encouraged us to find what we were good at and what makes us happy and then, if we needed help, they were there. They were the type like, 'You could do it. Go do it.' So, we had to be independent and figure things out."

With His Excellency Anthony Thomas Aquinas Carmona, the fifth President of the Republic of Trinidad and Tobago.

There is a direct link[4] between that kind of parenting and the ability to develop self-regulation skills and internal security, and to look ahead positively to the longer term, even when things feel messy right now – exactly the qualities needed by a woman aiming for longevity and success in the demanding and often volatile oil and gas industry.

"In my moments of mum-guilt, I sometimes lean on my childhood to reassure myself that I'm raising independent children. They know Mum is going to be there for all the important things, but there are some things I'm just not going to be there for."

Corporate vs Technical – Where Women Sit

From another angle, organisational structure might well be the key determinant of Giselle's success. A recent report commissioned by the World Petroleum Council[5] notes that within corporate functions, i.e., business and administration roles, women are more highly concentrated than any other functional area in the industry.

Giselle is aware of her positional privilege within BP and acknowledges it candidly. "I have not personally experienced a lot of the trials women face in the industry, but it's because of where I sit in the organisation and my role. If I look around on my own leadership table – HR, Procurement, Legal – there are females there all along. It was very rare, however, to find a female Project Manager or Operations Leader – those technical roles. I find we are seeing more of them now, though, and that's great."

When pressed to explain why female leadership is so rare in the technical disciplines, Giselle cites the family sacrifice involved in a wife and mother having to work offshore or be onsite. She also mentions the difficulty in rising through the ranks because numerically there's more competition than in the office-based roles, and fewer women in the pipeline to compete.

She also admits that sexism is much more obvious, and it creates a much more challenging work environment – at least, so she hears from women in the company – across the less-senior levels of the industry and in the technical areas, i.e., the field, the offshore world, and on the plants.

"I have heard females say that they have been told by men, 'I don't take instructions from no woman.' Or sometimes their actions are less overt such as the men not wanting to sit at the same table as the women in the galley. Sometimes, that chauvinism is a fear-based reaction in the men. The offshore world has been male for so long, it is difficult to adapt behaviours when females enter their environment. Throughout my years in the industry, I have seen this improve, though, as more women are taking up site roles."

"And what about the tables at which you sit? Boardroom tables, C-suite tables? Have you encountered chauvinism there?" I ask.

"In my own organisation, I don't see a lot of open chauvinism. Externally, I've seen more of it. In the very early stages, I remember being in government boardrooms where I was one of very few females, and feeling like, perhaps, my contributions were not valued. However, we have come a long way and now many of the most senior officeholders in government ministries are women."

Giselle admits, though, she has felt that, as a woman, she often must work harder to get her voice in the room, and to get her views heard and valued. "There were times where I may have felt sort of dismissed, like, 'Alright, we hear you, but on to the next.' " However, with time and experience, she has learned to become assertive and meaningful in her contributions to make sure her views are heard.

Here again, her membership in female-centred professional organisations has provided invaluable guidance. One of the most useful tips she's received is

With the Juniper platform.

about lobbying and allyship, i.e., the fewer women in the room, the more important it is to throw their voices in support of each other. So, if she's going to a meeting where she plans on raising something contentious, Giselle has learnt that it's best to "socialise" her ideas with the other women on the team beforehand – sometimes, even certain men. "So, they know what I'm going to say, and I know they're going to back me up. And, just generally, if one of my female colleagues is making a valuable point about something, I make sure to reinforce and articulate my agreement."

Key Responsibilities and Skill Sets

In 2015, Giselle's role was broadened to the present title, VP Corporate Operations. So, what does she actually do on a day-to-day basis?

Giselle explains: the level above her is the President, the country head for Trinidad and Tobago. Below him/her sit all the VPs or heads of various entities, and together they form the leadership team that develops the strategy and integrated plan for Trinidad. Much of her job involves supporting the President and his/her engagements with stakeholders or with staff. However, BP being a multinational, she also reports to people abroad. Then there's the matter of engagement with business partners, embassies, government departments, Chambers of Commerce, NGO groups. Some days are spent in the Mayaro community – BP's "host community" which houses the company's onshore operations facilities – engaging with community stakeholders. There's quite a lot of after-hours work; there are always events and meetings that run late or start very early.

Giselle obviously does a lot of "people-ing" – which can't be easy. But she says, "I consider myself an ambivert – an extrovert and introvert – so I'm not fearful of being in an external space. I do thrive on this and get a lot of fulfilment from engaging with various folks. But it takes a lot of my energy. So, there is a part of me, the introvert part, that sometimes just needs the quiet time to recharge."

When I ask Giselle to identify the key skill set she brings to her role, the key reason she has succeeded in it for so long, her answer is simple. "In living out my personal mantra of stay curious and add value, my superiors see me as: Giselle, who helps to solve problems or move impediments."

That makes sense, in an industry typified by all sorts of challenges.

She explains that many times, in delivering energy projects, it takes more than the engineering and design and the technical risk management. Very often the wildcards lie in areas of non-technical risk: Why is this regulatory approval taking so long? Is the host community informed or happy or not happy about this? Do we have issues with any stakeholders?

It's these types of external risk that can shut a project down. For every day of non-productive time in oil and gas – whether it be on a project or drilling rig – hundreds of thousands or millions of dollars are lost. And that's where Giselle and her team bring value. They bring the access and connection with the wider world which is crucial to keep things moving forward.

Sometimes, project teams may be so internally focused on a project's technical requirements that they are unaware of what's happening in the external world. It's Giselle's job to "read the room", i.e., the wider environment in which BP operates, with a view to understanding and flagging those risks.

"It's not easy to get people to see what they can't see, or to deliver the bad news. But it's important to do so in a solutions-oriented way, saying, 'This is the risk, and this is likely to happen. These are some of the challenges that we're seeing here, but here's how we can resolve them.'"

"I consider myself an ambivert, so I'm not fearful of being in an external space."

Do Qualifications Matter?

Why did Giselle, after ascending into senior leadership without it, decide to go back for her MBA? "Did they make you do it?" I ask, meaning her employer, BP.

"No," she says, "I never felt that I needed an MBA for my next promotion. The company I work for really values experience and what you bring to the table over any sort of credentials. Particularly in my field. So, there was never any question from the company."

Instead, she shares about waking up one January morning in 2017 and just deciding that was the year she was going to do it. By then, her kids had become more self-sufficient and the whole MBA market had changed. There were now reputable low-residency programmes available, which combined a small residential component with mostly online instruction, allowing her the flexibility she needed in her busy life. "I did my MBA with the University of Cumbria in the UK. I'm very grateful that I went back to that life goal I had set for myself."

"Do you think we women sometimes feel we need to bulk up on credentials in order to sit in a room?" I ask.

"I think there was a little bit of that," she admits. "I was also getting appointed to sit on external corporate boards. I did have moments of wondering

Receiving UWI's Distinguished Alumni Award

if I was enough, if I needed more of something, to be a good Director. But besides those moments of self-doubt, pursuing my Masters was more about staying curious and continuously learning."

When Giselle says, in a celebratory tone, "It took me twenty years, but I did it!", I ask if she'd name that as her greatest triumph.

"No," she replies. "The greatest triumph of my career has been all the moments that I've been instrumental in putting Trinidad and Tobago on the map, globally. Those moments where I have helped to get investment to come here, to get ideas to come here, to get energy or sustainability projects to come here. Not saying I do that by myself – it's a team effort – but I know I have played my part in pushing to make sure little Trinidad gets the attention that's sometimes hard to get in a huge multinational business operating in close to eighty countries."

Leading and Managing as a Woman

Teams, teams. She is always careful to mention her team. So, I enquire about the team dynamics, from her perspective, asking whether as a female team leader she has perceived a difference when managing men versus other women.

In her answer, Giselle is at pains to avoid overgeneralisation, pointing out that every employee is an individual "with different needs and drivers". Some need more listening, more face time, more space to express themselves, and more reassurance that they are doing the right thing. "Others are more self-assured and like the space to be self-directing. Leading a team is about getting the best out of your people and knowing how to adapt your leadership style is important," she says.

Giselle agrees that many of the sensitivities and skills developed and deployed in being a woman, a wife, and a mother – collaboration, consensus building, empowering disparate personalities – are transferrable to being a team leader in the workplace. Again, while acknowledging the risk of overgeneralisation, she explains that from her observation, male leaders tend to adopt a more directive style. They are probably stronger at articulating a vision, they drive a harder performance edge, but may be less comfortable with conflict or emotional situations, and less empathetic: knowing their people less – or caring to know their people less – than female leaders. "Of course, throughout my career I have also had the privilege of working with men who genuinely cared about their people and were deeply invested in their development and success – I am a product of that," Giselle concedes.

Confronting Unconscious Bias

"What you just described in managing women, do you think that might be part of the reason why fewer women tend to get ahead? Because their bosses might be males who don't have the insight or the tools or simply don't see the value in adjusting their management style to accommodate the dynamics presented by female employees?" I ask.

"I have no doubt that does play a role sometimes in women not getting ahead. There are unconscious biases operating in how recruitment or promotion decisions are made. I have seen those play off in our industry, and although I don't think it's intentional, it really angers me."

Sometimes, unconscious bias comes clothed as a kind of well-meaning paternalism, where senior men are making choices for female candidates, instead of letting the female decide whether she's willing and able or not, e.g., "I can't offer her the job because it's an offshore role and she has four kids." But, as Giselle argues, this only begs the question, "So, the guy you are hiring doesn't have kids as well?"

Giselle's own career trajectory has been a testament to the fact that, for a female employee, so much turns on the mindset of your manager. I enquire what, if anything, her company is doing to address the issue of unconscious bias, which she may have avoided but which, by her own observation, is still alive and pumping.

"We have a very active DE&I agenda," she replies, then proceeds to elucidate BP's Diversity, Equity, and Inclusion goals. "We know we're doing great in the

onshore roles in terms of gender equity. And we've put deliberate programmes in place to address the challenges we still see in the offshore and onsite spaces."

Giselle also espouses the benefits of having a dedicated coaching and mentorship programme for females.

"Sometimes, when you're in a challenging time – whether it's family or work – you think dropping out is the only solution. Often women are afraid to ask for help because they think it looks like weakness. Or they're afraid to ask for accommodation. So having those types of programmes in place helps them to see there are other solutions."

Coaching programmes can help empower the employee to speak up and present her case to her manager, can help prepare an employee for job interviews if she's in line for a promotion, or if her present role – e.g., an offshore posting isn't working out, coaching can help her uncover different career pathways within the organisation.

"So, I think it's about giving them a safe space to be vulnerable and get the support that they need. And the same for men, right? They need safe spaces too."

Above all, when it comes to DE&I, Giselle is a firm believer in targets because, she says, "They push you to do things differently, to take action, and to measure progress. Not only to mentor and support differently, but also to recruit differently and build your talent pool. She makes reference to the statistics from our local universities, which show females graduates far outnumbering males. "So why are they not making their way through?" she asks.

BP is trying to address this imbalance by interrogating its own recruitment and promotion practices: "Is the candidate slate diverse, or are all of the candidates men? Do we have a good mix of candidates both in terms of gender and ethnicity? Who sits on the recruitment panel? Are they representative of more than just the HR group? Are there females on the hiring panels to counteract those inherent biases that may have been built into the system over time?"

I ask Giselle if she truly believes these DE&I initiatives are making an impact. She swears they are and predicts the gains will compound and accelerate over time. "We'll get there because we have firm targets, we are taking action to level the playing field and we are measuring progress, and we are seeing improvement."

"So, if your daughter says she wants to enter the oil and gas industry, in a technical role, you'd be okay with that?" I ask.

"I would hugely encourage that. In this energy transition that's upon us, the industry is undergoing rapid evolution, so not only do we need traditional oil and gas skills, but we need new skills as well. We're looking at renewable energy. We're looking at impacts on environment, electrification, data science, robotics. And the industry has always been on the cutting edge of innovation and technology development, so this is a wonderful industry to learn and grow in. There is no limit to the breadth of skill and talent we need. The energy industry is not a small world anymore."

Impact on Her Personal Life

At this point, we revisit the earlier discussion on role modelling as I enquire whether her daughters have expressed an ambition to emulate her.

She recalls there was a time when her eldest daughter was fond of saying, "Mum, I want to be just like you when I grow up. You can dress every day in your suits and your heels. And you work for a great company." But now, that little girl is eighteen, on her way to university and Giselle has taken great pains to impress upon her that the corporate model of what a successful female looks like is not the only model.

"I try to explain that they shouldn't limit themselves. This is the old model of success. It's what we knew: come out of school, get a job, go work for someone. But now, in this whole new economy with hustle culture, there's just so much opportunity. Success doesn't need to happen in heels and suits, it could be jeans and boots."

And is her husband proud? Giselle says, "Yes, he's very proud. And I think sometimes we underestimate that your life partner plays such an important role in your success. He's always been a cheerleader, a supporter, and willing to do his part in terms of child-rearing and managing the house, without resentment. That support or lack thereof could make or break you, as a working woman."

Finally, we get to the core of the matter, i.e., whether Giselle is proud of who she's become in this industry. She replies, "Yes, and I think I've grown a lot. Some of it is just maturity, but some of it is what I have been exposed to in the industry."

With husband Paul Thompson and her daughters.

It's an interesting irony to hear Giselle explain that even as her personality has been changing, her backbone hardening in a heated industry replete with men and Type-A dominant folks, there has occurred a simultaneous deepening in her consciousness of the fragility of our world.

"My job has given me the opportunity to sit much closer to the fire that drives the future economic prosperity of our country. My job is to understand risks – whether they're environmental or social – and how oil and gas affects people and how we mitigate those risks. It has caused me to develop a consciousness, awareness, and commitment to sustainability. This is my country, my home. My family lives here. Everything we do impacts people and families, positively or negatively. It's a big world, but we are so interconnected." ■

Endnotes

1 Yanosek, K.; Abramson, D.; Ahmad, S. 2019. McKinsey & Company. "How women can help fill the oil and gas industry's talent gap". https://www.mckinsey.com/industries/oil-and-gas/our-insights/how-women-can-help-fill-the-oil-and-gas-industrys-talent-gap

2 Nobel, Carmen. May 15, 2015. Harvard Business School Working Knowledge. "Kids Benefit From Having a Working Mom". https://hbswk.hbs.edu/item/kids-benefit-from-having-a-working-mom

3 The authoritarian style has been attributed to historical influences including slavery, colonialism, and conservative Christian religion. See: Burke, T.; Kuczynski, L. 2018. "Jamaican Mothers' Perceptions of Children's Strategies for Resisting Parental Rules and Requests". *Frontiers in Psychology* Vol. 9. doi: 10.3389/fpsyg.2018.01786

4 Dewar, Gwen. 2010–2023. *Parenting Science*. "The authoritative parenting style: An evidence-based guide". https://parentingscience.com/authoritative-parenting-style/

5 Von Lonski, U.; Syth, A.; Trench, S.; Goydan, P.; Riemer, P.; Fjaeran, T.; Miras, P.; Merchant, W.; Gauthier-Watson, C. 2021. A collaboration between the World Petroleum Council and Boston Consulting Group. "Untapped Reserves 2.0: Driving Gender Balance in Oil and Gas". https://www.bcg.com/publications/2021/gender-diversity-in-oil-gas-industry

CHAPTER 4

DEBORAH BENJAMIN

The Quadruple Burden: A 25-year Show of Strength

MY NAME IS DEBORAH BENJAMIN, I'm forty-seven years old, the mother of a seventeen-year-old son, and I'm currently Managing Director for ASCO, the global energy logistics provider. In 2009, I had the honour of being named on "the eList" as one of the Caribbean's most powerful women. Truly, there are few women in senior leadership positions in my field. However, the path to success has not been an easy one. To paraphrase a popular song from my teenage years, many will see the glamour and the glitter and assume a bed of roses. But not many will know that before making it onto that eList, I passed Big Mac combos through the window where I reigned as "Drive-thru Queen" as I worked part time while studying for A-Levels – that's how I funded my schoolbooks. And before that, while preparing for O-Levels at sixteen years old – always looking for something to keep me financially independent – I was the girl who received Betamax and VHS tapes at a little neighbourhood video store and sat in the backroom all day rewinding.

I like to work. I don't think I will ever not be working. I learnt the value of economic independence from watching my mother. She was a career teacher who, for many years, bore an abusive marriage with silent dignity, and then found the courage to transition herself and her children out of it. I saw what work did for her. Once she was able to escape the mental prison, it was work

Deborah Benjamin

and financial independence that allowed her to stand on her own two feet and mind us. That example drove me to be a self-starter.

My first "big-people" job, with an airline, allowed the flexibility and income to put myself through university. It also developed my ability to face a myriad of clients and client personalities, while managing and delivering on their expectations.

I believe that's been one of the greatest strengths propelling me through my career, which has since taken a unique flow from downstream retail petroleum products and industrial gases, into distribution, and more recently, into upstream energy services. There was some segue into retail industries, which showed me the value of making every bit of small margin count, but ultimately, the oil and gas sector has always pulled me back in.

Entry into the Oil & Gas Industry

After my airline experience, other client-focused roles in marketing caused me to move to the Eastern Caribbean (EC) island of St Kitts, where, on the recommendation of a customer, I was interviewed for a position with Shell, the international oil and gas giant. It was a lengthy multi-country interview process, which ended with me receiving an offer to become a Commercial Executive for the North EC region.

Admittedly, I was a little overawed by the multinational "big name", and concerned as to whether my non-engineering background would pose a liability. However, I quickly understood that what Shell needed was somebody to do the work which can only be accomplished in small, one-on-one interactions. They wanted somebody who's the face for the customer. An interface between all the technical guys and the end user. Someone who knew how to talk to non-technical people, and work the accounts, and listen to the customer, and bring what the customer wants back to the company, and just manage the client relationship.

And wasn't that exactly my skill set? Honed from my days behind the fast-food counter, deepened during years of appeasing irritable travellers.

But I knew I'd have a lot to learn. In order to do what I do best – manage and service the customer – I needed to learn the core business. I had to understand how the operations worked. So, I went to the guys very humbly. I sat down with them and let them teach me what it took or what they needed from me to support the business I was bringing in.

That's the first "burden", I think, for anyone entering the oil and gas industry from a non-technical background: learning the job. The industry is not only price-volatile but also dynamic in terms of technological change and even corporate linkages. Learning will always be part of the work. In retrospect, I think I've been successful across so many countries in the Caribbean oil and gas industry, because anywhere I go, I make myself a sponge: I sit, I absorb, I learn. I don't assume that I know it all. I'm never too high in the organisation to learn.

Ascension to Senior Management

So, I was in that Commercial Executive role for several years, managing most of the accounts of the Northeastern Caribbean – Antigua, St Kitts & Nevis, Anguilla, St Maarten, and the British Virgin Islands – then in 2005, Sol Petroleum acquired Shell's distribution operations in the entire Caribbean. The exit of several expatriates created a favourable environment for me to ascend further, based on the breadth of my regional experience. At twenty-eight years old, I was appointed to a Country Manager role, becoming one of the first female country heads for the multinational. Here was another daunting opportunity: to be a woman in oil and gas, and be that young, and be asked to lead – it could have proven too much. However, I was determined to give my all to the role. And I did, sometimes, to the detriment of my personal life.

The literature about women dropping out of oil and gas always mentions the weight of the "double-burden", i.e., the paid day-to-day job plus the unpaid work of caretaking a family at home. That's absolutely true.

My greatest triumph in the oil and gas industry isn't some corporate sales or managerial benchmark, but rather, the fact that even as a single mother, I've been able to endure that double-burden. I wouldn't say I had no help, but having mostly worked in foreign countries and away from my own family support system, I had to pay for the help I got. And I had to work thrice as hard as the men who were my colleagues. They all had wives taking care of their kids.

Single motherhood is hard. I will never win "Mother of the Year". I forgot my child in day care already. I was working, had my phone on silent, and

Receiving two awards from Sol in 2010.

then before I knew it, it was 8:00 pm and I had what seemed like ten thousand missed calls. I felt terrible. Like the worst person in the world. However, I've learnt over time that kids survive. My son has survived and is in his seventeenth year. And he's perfect. He's gone through his issues but I'm confident he will find his way in the world. And I'm putting a lot of faith that he will remain true to his mummy...but I'm also not holding my breath.

Present Role and Management Style

After several years in that Country Manager role, I went on to hold senior leadership positions in other oil and gas companies across the Caribbean. For example, in Jamaica, I worked for an industrial gases company; in Trinidad, I was NP's General Manager of Operations Fuels, with responsibility for the Retail Gas Station network, Distribution, Operations and Maintenance.

Currently, as the Managing Director of an upstream, multinational logistics company that supports drilling and production platforms here in the Southern Caribbean, I continue to lead from a perspective that draws heavily upon my foundational customer-engagement background.

Sometimes, the "customers" are external and sometimes they're internal, e.g., employees. Regardless, diplomacy and humbleness must go into it, to achieve an understanding with the customer. The person must be allowed to say what they are experiencing. It takes a lot of time and emotional effort to absorb the energy and feelings of others, and to ensure that even if I cannot give the response they're looking for, my "No" has still left them feeling valued.

I think I have my son to thank, partly, for improving my management style. Being a mother dealing with his unique challenges has forced me to develop patience, and I think that made me a better manager than I was before, because face it, whether you have children or not, you are called to "mother" at work all the time. Whether it's colleagues or subordinates, the ability to nurture is what creates relationships, and a strong fabric of relationships is what, in turn, gives an organisation its strength.

I haven't had female bosses, due to the side of the industry I've been working in, but I have had female colleagues on the same level, and I've witnessed how they infuse their leadership roles with empathy and emotional intelligence. I believe those are traits which most women leaders bring to decision-making.

I think another key trait that women bring to leadership is our nose, our intuition. We can smell a rat. We have this innate sense of knowing where to go and who to listen to. You just see things and you sense a dynamic that's happening. Quite often, supervisors and middle management will not disclose things that are happening on the ground, partly because they fear self-incrimination. This allows rumblings of discontent to begin and then snowball into crises which could have been averted, if addressed earlier. So, my management style is to engage directly. I'm not somebody who stays in an office. I come out to see what you're doing. I talk with everyone, at all levels. I ask each person: "Do you have what you need to do your job? How are you? Are you working safely?" By being on the ground and engaging, I can hear the rumblings and deal with them. Or at least understand, firsthand, where something is coming from and bring it up at senior management meetings so we can resolve the issue.

Leading and Managing in a Gendered Environment

Without a doubt, the industry is male dominated. It's most glaring on the technical and operations side, and at the uppermost decision-making level – my roles have always spanned both. Countless times, I've travelled to conferences and meetings, looked around and seen only men. It can feel lonely, at times. I haven't had bad experiences managing men as subordinates. Rather, most of the gendered conflict I've experienced has come from men who are on the same or higher seniority level as myself. Those are the people who typically try to challenge my competency and authority. It's as if they believe I'm occupying a space which should be held by a man.

To deal with this, I make it my business to know what's going on in the company, so I know your job better than *you* know your job. So, when that time comes and you do decide to challenge, I can answer you the way I'm supposed to answer. Not on an emotional basis but on a factual, technical basis. Emotion is what misogynists want to see. Every time I have been challenged by a male, it has been as part of a deliberate ploy to elicit and weaponise my own emotions against me. They want to see you hang up. They want to see you get scared. They want to see you break down. But when you challenge them back technically, they don't know if to move left or right after that.

Of course, this has added an extra layer of preparatory work to my managerial roles. You could say, it's the third burden: overpreparation. That's a weight from which, I believe, men are exempt. They show up and just wing it. But women can't afford to wing it. Because the moment somebody realises that you're winging it, they will embarrass you.

In this industry, it can be hard for women in leadership roles to trust themselves to just show up and be enough. I haven't always trusted myself. I think this industry can sometimes be guilty of gaslighting women – no pun intended. Repeatedly, I've been headhunted based on my résumé, seen my hiring publicised as a triumph for diversity in the C-suite, then, as I've settled into my role and tried to bring my unique expertise to the company's decision-making, I've encountered the expectation that I would just be an invertebrate who would sit quietly and tow the corporate line. At times, there was also the insinuation that I should be happy just to even be in the room. So, yes, sometimes I've doubted myself and even doubted the bona fides of my recruitment: *Why did they hire me? Am I here to just fill up a quota?*

Ultimately, though, I've never been put out of a job. I've always left because of some kind of misogynistic behaviour. They make you feel less-than. Your competency is being questioned, you feel dismissed, your reputation is maligned. You feel trapped. At a certain point, when you're boxed in, you do think about walking away from the whole industry.

I've been told, in a boardroom, that I should let my boss do all the talking in a meeting and that my job was to "just sit down there and look pretty". I've also been told by a superior in the industry that I would never be able to do my job well because I am a woman. One particular boss pushed me very close to a mental breakdown. I thought I was so completely incompetent and that maybe he was so right. And how can I even be in this role if I am that dumb and I cannot function? I was no good to myself and I just kept feeling that I was less than a one-cent piece.

Then a breakthrough came, at my lowest, in a moment where I recognised that my situation at work was a more sophisticated form of the abuse that I had witnessed at home, growing up. The problem, I realised, was not incompetence,

but incarceration. They put you in a mental prison that you must escape. But nobody can bring you out of that. You have to be ready, for yourself, to say now is the time. And it was through the support of my "ride-or-die" tribe of women friends that I was able to build the mental strength and emerge from that dark space.

Turning Personal Lows into Lessons for Others

That hurtful period in my early oil and gas career operated as a sort of initiation – a rite of passage, I think – from which I emerged confident and way tougher than I had been before. The deeper insights I'd obtained about the "mental prison" and paralysis many women face in abusive situations propelled me, in later years, to join the Family Support Network in the British Virgin Islands, where I not only served as a Director, but also supported many women by sharing my own story. I still passionately advocate for the advancement of women, and I'm now regularly engaged, across the Caribbean, to speak on the subject.

In fact, after twenty-five years in oil and gas, I have no hesitation in saying that the high points of my career were those moments when I was able to use my platform to invest in people. Working with families, supporting youth, providing mentorship and internship opportunities to young women, getting involved with youth athletics – sometimes even running on the same track with the kids and travelling, at my own expense, to see them compete across the Caribbean. The personal joy and satisfaction it gave me was the greatest return on investment.

Even now, when I see certain names coming up with a World record for their participation in the Olympics, my reaction is, "Oh my God, I was there when they were twelve!" So, I go on their Facebook page and show support and they remember me and say, "Miss B, Miss B!"

Her Mother as Role Model

My mother has been my main role model, although I've been fortunate enough to have other staunch supporters, like my Aunty Paula. Apart from work ethic, Mummy also taught me eloquence, diction, and poise: the ability to use words to convey authority, while remaining humble and respectful. She has never told anybody anything in a bad way. I guess maybe she developed that trait growing up in a Chinese household where respect for everybody and the concept of allowing people to save face was important.

The main lesson Mummy taught me, the one I rely upon most heavily in my work life, is: Don't do anything to anybody that you don't want done to yourself. I've had the unpleasant task of sending people home, whether by virtue of disciplinary action or retrenchment. But the first thing I ask myself is, "How would I, if I was in this situation, want to be treated?" I don't just hand somebody a letter. I still spend time sitting down and explaining to them, trying to get them to a "come-to-Jesus moment" to understand. And I don't think I've ever had a situation where anybody has not been able to understand. I think sometimes employees act out because they feel decisions are imposed, without bringing them into the loop, so they feel as if they've been cutoff from the hip – I try never to do that.

From Mummy, I also learnt that it's important not to be a stoic mother. It is important to let our kids know that we have struggled, that we have made mistakes, that success has "cost us plenty". The point is not to saddle them with guilt, but to teach them resilience, and hopefully, gratitude. I try to model that approach with my son.

The Need for Self-Care and Boundaries

As a child

While the record shows I can make and implement the hard corporate decisions, I must admit it does drain me emotionally, because people are not just numbers on a financial report. It weighs on me to know the position I'm putting you in: that you don't have a pay-cheque coming next month and you have kids to mind, your house to pay for, etc. So, there are seasons when I've found it urgent to respect my own mental-health limits and to withdraw into a regenerative shell. I spend a lot of time at home by myself. Because that's how I'm able to cope. I've also found solace in learning to play our national instrument, the steelpan. I'm a member of the Trinidad and Tobago Chinese Steel Ensemble, where I play the tenor pan as part of the full orchestra.

As a single woman in a male-dominated industry, self-care has also meant being purposeful about drawing physical boundaries. That's perhaps the fourth burden: devising strategies for those inevitable moments when men try to impose themselves. It's an occupational hazard. I've had to develop practical techniques to preserve my personal space and to physically extricate myself from compromised positions. Part of the job is socialising and attending events, so I've had to set parameters and be very deliberate about how I manage encounters. I only do dinners if it's a group. Every other time, if it's a one-on-one, I'll do breakfast or lunch. I never let anybody drive me home or back to the hotel unless it's in a group. Or I take a taxi.

Not only must a woman prepare for unwelcome overtures, proactively envision routes of escape, and extricate herself from situations where her trust is breached, but, quite often, we must find a way to do so while preserving the trust of the male colleague or boss making the advance. Otherwise, it changes the whole relationship dynamic. If you make them feel like an act of sexual harassment has occurred, they are likely to retaliate because now you're a threat

My link to the future, making it all so worth it: my son Shawn.

who may potentially damage their career. Remember, if it's a boss, they have to do your performance evaluation. If it's a male colleague who you must depend on for projects, they could sabotage your work.

Thoughts on DE&I Initiatives

The big companies in the industry have put a lot of effort into Diversity, Equity, and Inclusion (DE&I) initiatives aimed at improving women's experience in oil and gas. In fact, nowadays, DE&I is becoming a key stakeholder metric in the industry. However, there are some men who feel disenfranchised by the changes. They resent the fact that so many women are coming into executive roles and coming into what they perceive as "their space". They feel like these women are being forced on them. So, they may put up a DE&I front, but the attacks don't go away, they simply go deeper underground. In my opinion, any young woman coming into this industry needs to be aware of these reactionary sentiments, because she may have to manage them as part of the quadruple burden. She needs to think ahead, do research, and act strategically.

In fact, this is my advice to any woman entering this industry: learn everything you can about where you are and where you want to go. Use the skills that you were blessed with as a woman, and remember to support other women, not cut them down. Keep your eyes open for opportunities. Be a sponge. Have a Plan A, Plan B, and Plan C.

Above all, remember to take care of you. ■

CHAPTER 5

SATIRA MAHARAJ

Beware of the Self-Fulfilling Prophecy

A Conversation with Engineer and Gender Relations Scholar, Satira Maharaj[1]

ELF-FULFILLING PROPHECY, A process through which an originally false expectation leads to its own confirmation. In a self-fulfilling prophecy, an individual's expectations about another person or entity eventually result in the other person or entity acting in ways that confirm the expectations.[2]

Celeste: Tell me about yourself. What has your work been like in the oil and gas sector?

Satira: I'm a Mechanical Engineer with approximately ten years' experience in Trinidad's petrochemical sector. I started in 2009 as a graduate trainee at PETROTRIN, and worked there for about nine years, in the Turnaround Department mostly. A "turnaround" is what we colloquially call "a shutdown": taking plants out of service, performing major maintenance works, then getting them back up and running. I was involved in the planning aspect, particularly contracts, costing, budgets. After PETROTRIN closed, I worked at Yara for a little while, then Methanex. Then I started my PhD in Industrial Engineering at the Faculty of Engineering at UWI.

Satira Maharaj

Celeste: Who would you be interacting with most in that line of work?

Satira: Other engineers, planners, schedulers, Contractor employees.

Celeste: So, a mix of people. Any challenges being a female in that refinery-focused environment?

Satira: Yes, which anyone can guess given my PhD thesis: "A Gender Transformative Approach to Engineering in Trinidad and Tobago". The work environment for women in the oil and gas industry has changed somewhat, but yes, I've had challenges. Sometimes, subtle attitudes that may not overtly come across to everyone as "gendered" or "discriminatory", but they did have an impact on my working experience.

Celeste: Can you give an example?

Satira: When you're driving on a plant, your vehicle must have a spark arrestor and must be diesel-powered because gas-powered vehicles are more combustible. The vans we had when I was a trainee were manual vehicles. I have a manual license – all the trainees had – but one of the workers would say to me, "Oh, why you're driving that beat-up old van? We need to get a nice little automatic car for you."

It's something that comes across as innocent – just a joke – but he never said that to the male trainees. Also, if I were made to drive an automatic car, it would effectively prevent me from entering the plant, which is where a lot of my work happened.

I've been catcalled, or as we say, *sooted,* at work. Once, I even ended up peeing in a bucket because there were no restroom facilities for women.

There are certain assumptions made, in that kind of work environment, which are not often or easily challenged. When I started working, I would go into offices and see posters of women wearing virtually nothing, which didn't necessarily offend me, but if women are viewed as sexual objects in that environment, how would they then look upon a woman co-worker? There were some more serious incidents as well.

Celeste: Good word there: assumptions. Assumptions made by whom?

Satira: By the *people* working within the environment. We have grown up with some traditional ideas regarding what it means to be a man, what it means to be a woman, the kinds of things that men and women typically do. And those ideas translate into the workplace more often than we would like to think. So, someone might come into my office, which happened to be very close to the copier, and often they would assume I was a clerk or admin assistant or secretary. That was not an insult to me, per se, because the admin staff worked extremely hard, but it's always interesting when I get mistaken for the secretary as opposed to the engineer. It has to do with how we value work and how we apportion space. There's this concept that office work is easy, and that women are better at office work and detailed paperwork.

Celeste: Were you working mostly with men or was it a mixed team?

Satira: It was mostly men in the refinery. However, my group of trainees included three women. We moved through the system together and I believe that was helpful for us. We had other young women trainees coming after us, who also fell into our little friend-circle. I'm quite sure it was beneficial for them as well.

Celeste: Beneficial in what way? In having somebody you could relate to?

Satira: Yes, however, even that friend-group tended to be more conservative than I was. I tend to question things. So, things that I might see as not sitting quite right, most folks would not think twice about.

Celeste: Perhaps *you* were a little bit more aware or sensitive to the biases in the environment?

Satira: Definitely. I went to a Historically Black College, Florida A&M University, for my undergraduate degree, which opened my eyes to a lot of those kinds of biases – except this time it wasn't necessarily racial biases, but very gendered biases. I had worked for a year, straight out of university, with a Fortune 500 company in the US, whose attitude towards race, gender, etc., was very different from the environment in Trinidad at the time. So having undergone some of their training, I saw things a little bit differently from my engineering colleagues who came out of UWI.

Celeste: You were more attuned to noticing microaggressions in the environment.

Satira: Yes, exactly. And as I said, there were microaggressions and there was also some more serious stuff.

Celeste: So, female colleagues didn't match your level of awareness. Did it still help having those women?

Satira: Having that solidarity and company made it easier for us to be accommodated. I don't like the word "accommodated" in that sense, but we did have to be accommodated because one person can sometimes be more easily overlooked than a group.

Celeste: Would you say there is, in fact, a boys' club?

Satira: Yes.

Celeste: And it helps if there's a girls' club, however small – that's what I'm hearing you say.

Satira: Yes. Numbers matter, but they are not the only thing.

Celeste: Was female competitiveness ever a problem? Because you know what happens in minority groups: sometimes, to survive, members of the group decide they're going to defect, if you will.

Satira: Yes, because it's sometimes a negotiation: How do you survive in that kind of environment? Engineering itself has quite a competitive culture, which I see as linked to capitalism, to patriarchal masculinity, to our economic systems at large. Women who make it into engineering, they tend to be very academically qualified: top of their class in primary school, secondary school. Many of these women have that competitive drive and they've always been at the top or very close to the top. The group of women that I went through that part of my career with, we were all very smart, we all fit that kind of profile. So yes, sometimes there was that sense of competition.

I've seen where someone agreed that a situation was not fair, but rather than rock the boat, it worked in their interests to go along with the unfairness. I've seen that kind of thing happen many times. Back then, I might have been hurt by it. However, now I understand it was their way of surviving the competitive environment.

It's very interesting to see the way people, in particular men and women, compete with each other, i.e., with people of the same sex, and what they hope to get out of it.

Celeste: Over the course of *your* career, what would you say was the main goal of the female competition that you observed?

Satira: The main goal of female competition, I think, is to distinguish yourself. To get ahead, not so much in the early part of your career where you are more supportive of each other and everyone is learning together, but when you get to mid-career and you're looking to get to the senior level. It becomes more important to distinguish yourself from everyone around you because there are so few chances for advancement.

Celeste: Did you ever feel that there was a male competing with you or thinking that you were taking a space which should have been occupied by a man?

Satira: There was an incident when I was a trainee working just a few months, around the time when Kamla Persad-Bissessar was campaigning for general elections in 2010, or maybe had just won. An older gentleman came into the pump shop, saw me standing there, and he went on a whole rant. The part of his rant which stood out to me was, "Allyuh woman coming in here trying to take man work!"

Mind you, he was not an engineer. We had different skill sets. He was a technician, having worked most of his life doing a certain job, and I would not have been doing the same job that he was doing. But my presence was clearly threatening to him in some way.

It goes back to the question: Who do these jobs belong to? Who is by default entitled to these jobs? Which, in turn, goes back to our assumptions about what it means to be "male" or "female" in society. There are still certain ideas of what a man is supposed to do and what a woman is supposed to do, even though they're both called engineers or they're both called technicians. For example, working on the plant versus working in the office, working day versus night, doing physical versus more mental tasks, technical versus people-oriented work – all those assumptions still happen in the engineering environment.

Celeste: Did you ever feel like he – and people like him – were just lone wolves expressing their own personal bias? Was the environment one where colleagues would jump in and ask him to tone down or was the environment itself supportive of that kind of bias?

Satira: In that moment, there were some other male technicians around, and they did try to tone him down. However, he just kept on shouting at me until he tired himself out. I did appreciate the solidarity I received from the other technicians that day.

Celeste: Was that your worst moment as a woman in the industry? "You keep mentioning 'more serious stuff'. Please elaborate.

Satira: There was a time when I was being bullied repeatedly. Someone senior to me kept making "jokes" about domestic violence: advocating beating your wife, saying that he was going to put some "domestic violence" on me.

At the time, I was a volunteer with an organisation that assists survivors of domestic and sexual violence, and I personally know

Mangrove cleanup in the Orange Valley community.

people who have experienced those things. So, of course, his statements did not sit right with me. Although I asked him to stop saying these things, he didn't. I realised he was baiting me: he knew I would not stay quiet or laugh, I would engage him. So, he sought to provoke me.

The situation worsened to such an extent that whenever I saw him coming, I would find myself elsewhere, because I simply did not want to interact with him.

I thought long and hard about my course of action. I spoke to senior women. After an off-the-record conversation with HR, I decided to report it to my Manager, who agreed to talk to the employee. He had the conversation, came back to me, and said that I was being too sensitive. He brought my Superintendents to tell me the same. There was a forty-five-minute conversation, after which I ended up in the bathroom crying. He said people didn't like feeling that they had to walk on eggshells around me, and that I make a fuss about things.

But the things I was making a fuss about were things like him saying to the female trainees, "Don't wear jewellery on the plants, you don't need to advertise yourself."

Apparently, that was seen as me being too sensitive.

I don't think he understood the level of strength it takes to speak out about things like this, in hopes of making a difference. And if I was truly sensitive, then perhaps I would have done what several people suggested that I do: simply leave it alone, laugh and giggle, and try to make the best of a bad situation.

Celeste: You mentioned speaking to women senior to you, did you feel a sense of being mentored by them?

Satira: It didn't feel like that.

Celeste: Would that have made a difference? If you'd had a senior woman mentor for your role?

Satira: Perhaps, but there were very few women in senior leadership positions. I started in 2009, and around 2017 was the first time I saw a woman Acting Superintendent at my company and that was a big thing for me. I have never in oil and gas had a direct female manager.

Celeste: Do you think the type of manager or management style you're dealing with – your direct manager, your direct report – would have made a difference? If you had been lucky enough to have a different type of manager.

Satira: Absolutely. And I have been fortunate to have a couple of subsequent male managers whom I have respected and learnt from. I am still in contact with a couple of them today.

Celeste: Your manager at that time wasn't given any kind of diversity training?

Satira: I don't think so. At the time, companies didn't have diversity training, inclusivity training. That wasn't considered a big priority. It's only now starting to take hold. I saw some diversity training happen at

Yara, and my direct supervisor came back and said, "This is really interesting. It's something that we need to think more about."

He felt like he learnt a lot and it was really heartening to see that kind of response.

Typically, though, in the environment in which I worked, these kinds of training were seen as almost a day off, or something to get out of the way.

Celeste: Are those kinds of training made available all the way down the chain? For example, that technician who verbally attacked you, I wonder if somebody like him would ever have been given that kind of training.

Satira: We have a hierarchy of organisations in oil and gas. In the bigger multinationals with an international element, you are much more likely to find those kinds of diversity initiatives. They are typically considered the Client companies. However, a significant portion of the engineers who work in oil and gas in Trinidad, work as employees of small Contractor companies – sometimes on a three-month or six-month contract. That whole Contractor relationship is a big part of the problem: how people are employed and job security. If you're working for a Contractor, you're not going to have access to that kind of diversity training. In many small Contractor firms, you even have to provide your own safety boots, helmets, and other personal protective equipment – even though that's directly in contravention of the Occupational Safety & Health Act.

Celeste: In that Contractor setting, I imagine it must be hard to complain.

Satira: Yes, certainly. I have mostly worked for Client companies, which gave me more power. So, when I was *sooted* at work, I had the authority to pull the guy down – he was working on a height – and say, "Why did you do that?" I had the power to go to our Contractor and say, "Your employees need to behave a certain kind of way when onsite." But if you are a female working for a Contractor, chances are you're not going to have that power to address gendered behaviour.

When I was working for Contractors, they didn't want to pay my National Insurance contributions. I paid their portion. So, if you don't have the power to negotiate something mandated by law, how can you have the power to command respect?

Celeste: Yeah, imagine getting pregnant in that kind of environment, asking for maternity leave, and expecting to keep your job.

Satira: The lack of power also has to do with unionisation versus non-unionisation. Collective bargaining agreements, etc.

Celeste: Have you ever felt that your professional competence was questioned simply based on your gender?

Satira: Yes. When they needed someone to climb into a column and do checks or keep an eye on Contractors, they wouldn't choose women, because of the demanding physical component of that kind of work. Which goes back to the whole design of the column. Who was it designed to accommodate, right?

Very early on, I was part of a group of trainees – three females and one male – eager to participate in a turnaround. It was simple work: running back and forth, getting materials. But they didn't choose any of the women. Perhaps they felt that the male trainee might be more familiar with the kinds of materials to be used, although we were all trainees at the same level at that time, and very willing to learn. Perhaps it was the idea of him having to move between different spaces, walking, driving, carrying flanges or whatever materials.

That's the thing about "technical competence", it can be a self-fulfilling prophecy. Sometimes as a woman, you want to learn a variety of skills, but you are chosen for certain tasks and the men are chosen for certain other tasks, even though you're supposed to be on the same level. I've been tasked with contracts, budgeting, a lot of mental work, logistics. And I was pretty good at it. However, I was shepherded, so to speak, towards that aspect of the job, as opposed to the more technical aspects, e.g., design and calculations and on-the-ground supervision of jobs that involve welding and fabrication, pipe fitting, and column work.

This "shepherding" influences the kind of skills that we women develop, then people form certain impressions of us based on our "skill set".

Celeste: Is it a kind of misguided paternalism?

Satira: Oh yeah. Another example: they were looking for someone to do the night shift, and I said I was interested. A couple days later, a male colleague of mine, from a different department, was selected. I went to my Superintendent and said, "Was I given consideration? Is there anything I can do to get a chance like that?"

He said, "Oh, I didn't think you were serious about being considered."

I said, "Yes, I came to you, and I told you."

He said, "Well, isn't it against the law or something for women to work at night?"

It's true, once upon a time in our nation's history, there was The Employment of Women (Night Work) Act, and yes, it did prevent women from working at night in industrial undertakings. It has that paternalistic, protective assumption built into it – not overtly – but you have to understand all the things that were happening at the time of enactment, e.g., the Moyne Commission Report.

I did some research, then went back to him and said, "No, it's not illegal for women to work at night. That was repealed by the OSH Act 2004. Process Engineering has women assigned to the shutdown at night. So, I'd like to be considered seriously next time." But I never did get to be considered again because PETROTRIN closed down before another opportunity came up.

Celeste: So, he knew of the old law, but didn't know it had been modernised. Kind of like the old adage: *A little learning is a dangerous thing.*

Satira: It was just something in his mind. He was being protective, I think, because if a woman has to work at night, in that environment, there would be concerns about her safety and her security. Women are considered at-risk employees: at risk for getting raped, sexually violated. They are often told to work in pairs, so I was often teamed up with another female engineer. And again...self-fulfilling prophecy... if you have to use two female employees to do the job where one male can do it, then reason dictates you use the one. Whenever I would ask about night work, the kind of response I would get was, "That's hard work or that's not a nice work, you know. You eh go want that."

So, there were people feeling that they knew what was best for me, as opposed to me being allowed to make my own choices and be considered for job enrichment in the same way that my male colleagues would.

Celeste: To paraphrase one of the commentators I've read, you would think an industry that could find a way to drill so deep into the earth could find a way to keep a woman safe at night.

Satira: Exactly. But historically, there hasn't been a will to incorporate women into male-dominated industries in the Caribbean. Joan French[3] has written about the Moyne Commission Report and its relationship to women. That Report investigated the terrible conditions in the Caribbean in the 1930s, and one of the suggestions was not paying every working person a living wage, but rather pulling women out of the workforce and inculcating the idea of the right kind of family and the right kind of woman – the housewife – thereby leaving jobs available for the men, so more of them could, in turn, support families.

Celeste: Interesting. But whatever social engineering the colonial rulers intended, it did not pan out well for them. Caribbean women haven't retreated gracefully into domestic obscurity. Anyway, let's switch a little bit and talk about *you*. I'm sensing a very strong sense of self and a courageous spirit. You stand your ground. Do you think that comes from the way you were raised?

Satira: I've had strong women in my life. I know women who have been taken advantage of, and I particularly don't want that for the next generation of girls and boys coming up. That's one of the guiding considerations whenever I'm facing a tough decision: What's going to help the next generation?

I feel it's important to take a stand for the things I believe in. Taking a stand can come in different forms. For example, you asked about the boys' club earlier; navigating that club might be one way to take a stand or negotiate and change things. Sometimes "taking a stand" happens in a very subtle way. Sometimes it is having a conversation or asking a question, planting a seed. It doesn't always have to be some big moment.

I was taking a stand when I signed up for courses at the Institute for Gender and Development Studies at UWI. I wanted to learn and understand why these things happen in the workplace – and what I was learning explained my actual life – then I ended up with a second Master's Degree in Gender and Development Studies.

Celeste: So, your gendered experiences as a female engineer drove you to learning more about why these things happen, and precipitated another career outside oil and gas?

Satira: Yes, that's exactly it. My lived experience combined with the learning give me a whole new appreciation for different power structures in society and how we use power, who has access to power and takes advantage of it – whether we're talking about social categories like gender or race or sexuality, or even skin colour and religion. And these are not things that are easy to talk about in the Trinidadian society, but that's the kind of thing I lecture on these days.

Celeste: Tell me about key influences in your family.

Satira: My father is an engineer, so I had an interest in technical things from a very young age. My "Tanty Network" – my network of aunts – is incredible. They have endured very tough times. I have a couple uncles as well, who are in STEM fields, and some are engineers. I never felt that I could not be an engineer because I was a girl. Far from that. I was a smart child growing up and I had a lot of opportunities. I admit, I am privileged, and I am grateful for that. We always had parents who were interested in our education. My grandmother educated all of her girl-children – it's a story that we are all very proud of – in a time when that was typically not done in Indian homes. So, I am grateful for those influences in my life. They have certainly made a difference.

Celeste: How did your dad feel about you becoming an engineer?

Satira: He's proud. He worked at PETROTRIN for 33 1/3 years. He retired the year before I went in, so I had the benefit of his reputation going before me, which is a privilege in itself. Not everybody has that, so I do acknowledge that.

Celeste: However, you did have that, as you say, privilege of his name going before you and yet you still encountered what you encountered. So, imagine for the girl going in there just cold and unconnected. Did you ever confide in him some of these negative experiences?

Satira: Yes, yes, that was one of the big things that I went to him with, when I was trying to figure out what was the best course of action. I had his support. He didn't know many of the players involved, though.

Now, I know better how to navigate those kinds of situations. I'm very analytical and perhaps if something like that should happen again, I feel I'd be better in dealing with it. I understand where those folks are coming from. And I know why it's happening.

Celeste: So why is it happening?

Satira: Power. How different people access power, use power, what it means to have or not have power as a man or woman, or Indian, African, Chinese person, persons of different sexualities, hair textures, religions in our society, and how we navigate those.

Celeste: I remember reading about the marginalisation of the West Indian male and how they're getting more and more sidelined. So, I'm guessing the oil and gas industry, it's probably one of the last bastions; engineering itself, one of the last bastions of maleness.

Satira: That's exactly right. I had to read Errol Miller's entire book about male marginalisation[4] for one of my gender courses. Miller argues that women's accomplishments mainly resulted from men being held back, in the specific context of the Jamaican teaching service. Unfortunately, in the Caribbean, the accomplishments of women and girls are often framed in terms of opportunities being taken away from men and boys. Definitely, the engineering environment is considered one of the last "masculine" spaces. Stan Gray[5] has a great article on the "shop floor" and I think it's applicable to oil and gas and engineering as one of the last bastions where men can be patriarchal men and express that kind of masculinity in a way that is not allowed in other areas of society.

Celeste: Okay, here's a brass-tacks question. Have they started making safety gears that fit women properly?

Satira: There's a safety store that sells pink and purple hard hats and safety vests and shoes catered more to fit women. Which is nice, some women really do love pink and purple. But it's more nuanced than just having pink and purple.

Celeste: Yeah, that's a bit of a cliché.

Satira: Now, in terms of coveralls, unfortunately, our boobs still don't fit into coveralls. They are literally made to fit a man's body. They have a zip right in front of the private areas so they can go to the bathroom very easily.

When *I* need use the bathroom – I'm hoping there's a bathroom on site, I'm hoping I'm not on my period because there's not going to be a bin for me to dispose of anything – it means taking everything out of my pockets, undoing buttons, unzipping, tying the coveralls around my legs, doing a little hover over the seat. It's much more complicated for a woman navigating this male clothing. If it was a regular shirt and pants, that's not difficult to manoeuvre. But many sites require full coveralls because many men don't keep their shirts tucked into their pants.

I'm short – five foot one and a half inches. My mother, fortunately, is a seamstress, she has always had to hem my coveralls. The sleeves are a challenge because they are made with cuffs. The bust and waist fit is a challenge. My torso is short, so to get the fit around the bust area, I need to go for a larger size coverall, in which case the waist tends to hang down.

At five foot one, I usually have trouble finding coveralls, gloves, goggles and harnesses that fit me.

Sometimes these simple things can pose quite a challenge. Like gloves – everybody uses gloves. Everybody climbs on the plants, and if you're climbing you need gloves. However, the gloves often don't fit. They slip off and that can be a hazard. So, I always get extra-small-fit gloves, which aren't always in stock. Similarly, before I got my tested safety glasses, as a trainee you had to wear goggles. They were huge and kept falling off.

Same problem with harnesses. If you're climbing over six feet and you're not on a platform, you need to have a safety harness, in case you fall. But the harnesses, even the smallest ones where I worked, would only be effective if you weighed over 120 pounds. Some of the smaller women consistently were below 120 pounds. So those harnesses would not have been effective for their bodies. And if you don't have the safety equipment, then again, it's kind of that self-fulfilling prophecy.

Employers design and buy equipment for the majority or for the typical, for the norm, but that has a side effect of excluding those who do not fit into the norm.

I would go further and say that in the oil and gas industry, the physical plant, the tools, etc., they are historically designed for the typical male physique.

Celeste: Wait, you're saying the ergonomics of the oil and gas industry are gendered?

Satira: Yes. We typically design tools for upper-body strength, and men have more upper-body strength than women. Even things like how we design ladders.

Celeste: I thought a ladder is a ladder.

Satira: No, I'm not talking about the kind of A-frame ladder that we have in our homes or a regular eight-, ten-, twelve-foot ladder.

In the industry, ladders are vertical, like 30–60 feet long, and are typically built along the sides of columns, heaters, etc. As I said, I'm five foot one and a half inches; my stride is shorter than that of a man who's five foot eight. So, on a vertical ladder, due to my shorter stride, I have to put in more energy for my legs to climb. Another thing is that there's a cage around the ladder, the purpose of which is to provide a sense of security, and if you feel tired, you should be able to lean back against the cage and rest. My superintendent, who was about six feet, was able to do that. My back, though, could not reach the cage. To get even near it, I'd really have to stretch out and my centre of gravity would be very far from the ladder. So those cages are essentially non-functional for somebody like me.

Celeste: Well, Satira, you've certainly opened my eyes to many things. Thank you. My final question for today: What would you say to a young female engineer today who wants to work in oil and gas?

Satira: I would say that it is wonderful that she wants to work in oil and gas. She has likely worked hard to become a young engineer, and she should continue to work hard to pursue her dream. Unfortunately, the oil and gas industry can be tough for women, not because of any failing on their part, but because of gender stereotypes and conventional ways of thinking. I would remind her that things are evolving because of the excellent work done by the women who have gone before us, and let her know that she is continuing their legacy. I would also let her know that she can reach out to me should the need ever arise. ■

Endnotes

1 Satira Maharaj wears multiple hats. She is currently a PhD candidate in Industrial Engineering at the University of the West Indies. Her research focuses on gendered experiences and negotiations in the field of engineering in Trinidad and Tobago. She recently presented at the Institute of Industrial and Systems Engineers (IISE) 2023 Conference in New Orleans. Satira holds a Bachelor's Degree in Mechanical Engineering from Florida Agricultural and Mechanical University. She also holds Master's Degrees in both Engineering Asset Management, as well as Gender and Development Studies from the University of the West Indies. Her research interests include Plant Turnaround Management, Ergonomics, Development, as well as Women and Labour. In addition to her doctoral studies, Satira is a Research Assistant at the Institute for Gender and Development Studies, where she teaches undergraduate classes in Gender Studies. She participated in a Gender Impact Assessment of COVID-19 in Trinidad and Tobago, and was part of the Planning Committee for the International Women's Day March. Satira also teaches undergraduate courses in Gender Studies as a (Part-Time) Adjunct Lecturer in the Labour Studies Department at Cipriani College of Labour and Co-operative Studies. Satira has volunteered with Conflict Women, Ltd., and is a One Young World Ambassador for Trinidad and Tobago. She enjoys Latin dance, as part of the Salsa Vive Dance School. Her thoughts can be found at her blog called "Diary of an Ex Oil Worker".

2 Jussim, Lee. August 1, 2023. Encyclopaedia Britannica. "self-fulfilling prophecy". https://www.britannica.com/topic/self-fulfilling-prophecy. Accessed: August 29, 2023.

3 French, Joan. 1988. "Colonial Policy Towards Women after the 1938 Uprising: The Case of Jamaica". *Caribbean Quarterly* 34(3/4): 40.

4 Miller, Errol. *Marginalization of the Black Male: Insights from the Development of the Teaching Profession.* Institute of Social and Economic Research: University of the West Indies Mona, Kingston, Jamaica: 1986 first edition and second edition, 1994.

5 Gray, Stan. "Sharing the Shop Floor". From *Beyond Patriarchy: Essays by Men on Pleasure, Power, and Change.* Michael Kaufman, (ed). Oxford University Press, 1987. pp 216–234.

"The greatest obstacle is yourself and your belief in your abilities. That is your initial block. The challenges will not change despite who sits in the chair, it is up to you to have a focus on the results and not on who can do it differently. There have been multiple challenges to be honest, but with a team that is supportive, including those in your immediate sphere of influence, it can make a big difference. I also pray. A lot."

Mechelle Smith, General Manager (Ag),
National Petroleum Corporation, Barbados

Suriname

CHAPTER 6

MARNY DAAL

The Pioneer Woman

Her twenty-year quest to find Suriname's offshore oil

ARNY DAAL, THE RECENTLY-RETIRED Director of the Staatsolie Hydrocarbon Institute, recalls being in London several years ago, at an oil industry event, when she had the conversation that would change the fortunes of her country, Suriname (formerly Dutch Guiana), forever.

Tullow Oil had just been awarded Best Explorer in West Africa. She approached their representative to offer congratulations and was surprised that he actually knew of Suriname. He revealed that his company had inherited an oil claim in neighbouring French Guiana.

"Then he showed me the seismic of the Jubilee field in West Africa. I looked at the line, and even not being a petrophysicist, I was able to recognise that what he was showing me looked like what we had in Suriname. So, I went home very enthusiastic and told the guys we have to study West Africa's Jubilee field. I felt it was possible that field was mirrored on the other side of the Atlantic, like a cookie that breaks in two, because the Guinea Plateau and the Demerara Plateau were once joined."

Marny Daal

First Communion.

With her sisters.

With her parents Orwine Vogelland-Mehciz and Leendert Vogelland.

Signing PSC block 61, 2018.

The only question was whether the cookie broke before the oil was made or after. "If it was after, we're in the offshore oil business," Marny told her team. Twenty years later, after many initial disappointments, detours, and detractors, she believes that chance meeting wasn't just a case of being at the right time, in the right place, but rather it was the Universe doing what it does: helping you, if it sees you working hard.

And Marny has always worked hard for her country. Hers is a professional career marked by international intrigue, political machination, and even hunger crisis on the high seas off Suriname. You could say she's "The Pioneer Woman of Suriname's Petrochemical Progress".

Yet, Marny isn't an engineer or geoscientist, she is a lawyer by training. She didn't grow up dreaming of leading Suriname's search for oil; she dreamt of being a police officer, like her beloved father. But just as she was about to graduate from high school, there was a military coup which resulted in many police officers being imprisoned. On the advice of her father, Marny went into law. And it was as a young lawyer, at Suriname's Ministry of Labour, in the 1980s, when Marny received her first call to national service upon the world stage. It was a troubled time for Suriname. In the wake of the coup, more than a dozen opponents of military rule had been murdered. One of the victims was the leader of Suriname's largest trade union federation, and when Suriname's case was brought before a human rights tribunal of the International Labour Organisation (ILO) in Geneva, Switzerland, neither the new Permanent

Rig visit to Block 58.

Secretary at Suriname's Ministry of Labour nor any of his senior staff were willing to attend.

Enter Marny Daal. She was a twenty-six-year-old lawyer at the Ministry, and by her own description, "very innocent" back then, in 1987. It was she who was dispatched for questioning before the committee on human rights.

"I arrived and I was like, 'What the hell am I doing here?' I was really scared. The first week I was trying not to get lost in this big UN building and the second week I was just crying on the phone because I felt so alone and so scared."

The Ministry continued to send Marny every year, and after that terrifying initiation, her fear began to recede. There, in Geneva, she began to embrace her first experiences of the international arena: how people think, how people take decisions, and what it was like being a young woman in a room with mostly men. "I would always prepare thorough reports beforehand, to avoid additional questions from the committee," she recalls. And the effort was clearly valued, because when it was Suriname's turn to represent the Caribbean, it was decided that she should represent the region on the ILO Governing Body, the executive body of the International Labour Organisation, which met every three months.

"I really worked very hard to get Suriname in a better position and to get rid of the stain of the violation of trade union rights of 1982."

However, as is often the case with those who dare to venture far afield, while Marny was doing great things for her country on the international stage, she was not above the reach of small-town politics. Government Ministers were being changed on a regular basis, and in 1996, Marny, who had by then become a senior person at the Ministry of Labour, was on vacation with her young children, in the interior of Suriname, when the seventh Minister in her career was installed. "We didn't have cell phones back then. I didn't have a radio. So, apparently I made a false start with the new Minister because I wasn't present when he was introduced to the staff of the Ministry."

She came back from vacation to a tense atmosphere at work. As she recalls, "I was really fed up; again, we had somebody completely new who didn't understand what we were doing. I have an attitude, so it clashed."

Marny resigned. Shortly thereafter, she was approached by Staatsolie, the national oil company, to set up their legal department. "I was mad, so I said yes."

In 1997, she went to Staatsolie, not knowing anything about oil but just knowing that she had to get out of the Civil Service. "At the level I was functioning, they wanted you to become a puppet of the political system, and I did not want to do that."

As one can imagine, for someone like Marny who had seen action on the international front, the in-house legal work proved monotonous and uninspiring. "You got a bunch of contracts in the morning, and it wasn't gonna differ with the ones you got in the afternoon," she says. "So, I felt really trapped in a cage and wanted to move out."

Soon enough, though, Staatsolie presented a challenge in a way totally unconnected with Marny's substantive portfolio. In 1998, within six months of her joining the company, problems with the government started, a sale of Staatsolie was announced, and the CEO was dismissed. Privatisation would, of course, have meant other larger-scale retrenchment. Having worked for so long at the Ministry of Labour, Marny, leading the new legal department of the company, contributed to the counteroffensive against the government.

She recalls thinking, "I went from the frying pan into the fire." Convinced that she had nothing to lose, and that her involvement in the fight was as moral and ethical as her efforts at the ILO had been, she assumed a prominent role in efforts to ensure the national oil company remained "national", was not gobbled up by some foreign multinational, and that Surinamese jobs remained secure.

"We took on the barricades, we spoke to politicians, trade unions, to schools, to the university, in order to stop the sale. I was put on a list to be laid off – top five – because the government didn't like the way I was defending Staatsolie. But, as a result of our battles, the government lost the elections. So, the people in Staatsolie saw me as somebody who was loyal to the company and could help develop it further."

However, once privatisation was no longer a threat and the company returned to some normalcy, Marny, who found herself back at her boring contracts job, announced her intention to leave Staatsolie for a position at the local alumina company. Lured by the fact that they were the best paying employer in Suriname at that time, she went through the whole process of being accepted, doing physicals, etc., only to be told by confidential sources that mischief was afoot. "Somebody said, 'You're going to be the first female manager in this company. But there's a catch: they know that within five to six years the bauxite resources will be depleted, and they'll need to lay off all these people within the company."

No way. Not Marny. She hadn't fought for labour rights and battled government retrenchment just to turn around and become an instrument of private sector layoffs. So, when the CEO at Staatsolie approached with a competing offer to stay on and help the oil company "rediscover its institutional task", the career choice seemed obvious.

He reminded her that Staatsolie had started in 1980 as a vehicle for regulatory oversight of a Petroleum Sharing Contract (PSC) with Gulf Oil. The then-CEO went on to invite Gulf to further explore/develop an onshore discovery of petroleum – just a small find with no seismic data – which had been made a long time ago by another international oil company. When Gulf declined, Staatsolie had started its own commercial operations, from scratch (in contrast to Trinidad where the national oil company inherited the operations of previous international operators). By 1982, Staatsolie was producing approximately 200 barrels a day of sweet crude. So focused were they on building a company and establishing commercial operations, that their original regulatory duty was gradually sidelined. It took the government clash of 1998 to bring Staatsolie back to the realisation that if they were just a State oil company, any Minister could decide their fate and sell them off.

So, Marny's CEO wagered that the government's commitment to Staatsolie would deepen if they were able to resurrect their oversight function and were successful in luring other oil companies to Suriname. That was the job he was offering Marny ("just take care of the regulatory role") and it seemed, to her, the opportunity to strike out, and by the sheer force of her own will and intellect, and her own sense of right and wrong, build a new presence for her country on the international stage.

Drilling activities of the Maersk Developer offshore Suriname.

Building new things, meeting new people: she'd done these since childhood, as the eldest of her father's three girls. "He was very handy and if he needed to build something, I was always there to assist and do things with him. So, I have never seen a problem in doing things that aren't traditionally for women or girls." The family frequently moved around ethnically diverse Suriname, living in different districts among different types of people, because of her father's police work. This engendered in Marny a confidence that not only could she do anything, but she could also relate to anyone, without fear or favour. "I thought I could conquer anything. I always loved adventure and wished there was a circus I could join and see the world."

So, she and one employee, a secretary, set out to establish Staatsolie's newest department called "Petroleum Contracts", with responsibility for making assessments of the country's petroleum potential, promoting investment opportunities on international fora, inviting international oil companies to explore areas where Staatsolie was not itself conducting commercial drilling, negotiating and monitoring petroleum contracts with third parties, and acting as liaison between international oil companies (IOCs) and the Suriname government.

Immediately, Marny began searching for know-how to promote Suriname – which was, by then, producing 8000 barrels a day – in an international environment where few people were familiar with the country. She made it a priority to attend Trinidad's Energy Conference every year, where she touted the successes of her country's onshore oil industry, while inviting IOCs to explore the offshore waters. She also used those visits to learn as much as possible from the attendees, and even secured a one-week attachment to the Ministry of Energy in Trinidad, where she met "Mrs Inniss", who she names as the only female mentor of her oil and gas career.

Eventually, Marny and her team developed the confidence to take their mission on the road, to America and Europe, carrying with them, like an ace up the sleeve, a report of the United States Geological Survey in 2000, which identified the Guyana Basin as the second largest underexplored basin.

"Come and have a look offshore. Come and do studies. Give us a proposal," was her constant invitation to the multinational IOCs. Marny and her team did everything they could think of to entice investment: shared data, held bid rounds, did client surveys – it was a time of constant trial and error, she says. And what was needed was enthusiasm, optimism, and an energy to match her own. So, she made a point of accepting the younger people, coming straight from university, who nobody else wanted because of their lack of experience – after all, hadn't she been just like them, in her earliest days at the Ministry of Labour? "It made my life easier, because I was able to make them enthusiastic. Tell them we're just like Columbus and we're gonna find it, don't worry, we're gonna find the oil. So, we were a very close-knit group."

Over time, their Petroleum Contracts department grew and evolved into the Staatsolie Hydrocarbon Institute. Along the way, though, Marny recognised that she herself lacked experience – technical expertise – and that in order to lead her people, she would have to become technical. She immersed herself in geology books and tried to understand everything connected with exploration, not only by reading but by seizing every travel opportunity to talk with knowledgeable people.

Team building.

That eagerness to learn from others is what had helped to deepen her conversation with the Tullow Oil executive from a mere congratulations to an actual geology lesson about the broken-cookie potential of the Guinea/ Demerara Plateau. The meeting also prompted another breakthrough for Marny, as head of Petroleum Contracts. "We used to think that we only needed the Shells and the Exxons – the big companies – but I saw that, in West Africa, it was these small, nimble independents that developed Ghana, Côte d'Ivoire, etc. So, I said we're going to target them: the smaller outfits."

By risking a shift away from the big-monied IOCs, Marny revealed herself to be not just a hard-nosed international lawyer and self-taught drilling geologist, but a person who deeply understands the pioneer spirit, i.e., the immeasurable and indefatigable energy of an underdog with a dream. "The oil industry is very volatile. And every time the oil price drops, the exploration group is the first to be let go by big companies. These laid-off geologists and geoscientists are the type of people who set up independent oil companies, with the knowledge they have. They are small, so you deal with the owners of the company, the very people who have vested their energy and all their money into a single dream."

Marny partnered with them, the dreamers. Although later, when TotalEnergies, Shell, Exxon, and other big names came, the groundwork for exploration in Suriname had already been laid by the nimble independents. It was through them that Marny and her team were able to finally discover that yes, the cookie had broken, *after* the oil was made, and that Suriname was sitting atop large offshore reserves. But it proved to be a long, emotionally turbulent journey

Well inspection in the onshore swamp oil field.

for Marny. For almost two decades, people had rubbished her belief that there was oil in Suriname's deep water. When she'd first begun advocating for offshore exploration, she was often dismissed by her male colleagues: "No Marny, this is technical, you won't understand." When the first well came up dry during her tenure, she felt physically ill for a week. And, over the years, every time another drilling campaign proved unsuccessful, snide comments were flung her way.

But she had never doubted.

"Me and my team, we just kept trying and we had some very interesting experiences because I think we were too eager to find the oil, and also, we didn't think too much about the consequences of our actions. We just did what we thought was right, and it probably helped that we were even a little naïve."

She took it personally the first time a seismic survey was ever done in Suriname; she wanted to be there herself to see history made. Her recount of the adventure sounds like a mashup of biblical events and miracles. Marny, along with two female friends who also worked at Staatsolie – a Geologist and a Corporate Communications Specialist – hired a small boat to take them to the big ship on the high seas where the seismic survey was being conducted. They were "very happy going up, until the big waves" hit the boat. Nevertheless, the trio managed to get onto the ship, observe how the survey was done, and even take pictures for public relations. Having achieved all they'd come for, the women were eager to head back to Suriname's capital city of Paramaribo, because they hadn't eaten all day except for two chocolates given by the ship's cook. However, by then, it was midnight and the tide had changed, and there was now a substantial difference between the big ship they were on and the small boat.

"You had to be very careful with your step. Because of the waves, you could get caught between the small boat and the big boat. Nowadays, I would probably say stop and spend the night on the vessel, then leave tomorrow. But I didn't think like that, back then. I simply wasn't afraid."

The smallest woman went first, she made it across. Then Marny went, she made it across. Oh, but the last woman – the "big woman who'd received the chocolate bars"– when she stepped toward the small vessel, the big ship moved. The guys on the small boat had to grab and haul her in, otherwise she would have disappeared headfirst into the dark ocean.

"Looking back, having fun, and joking about it, we all could have died – just because we wanted to know firsthand what is a seismic survey, and be able to say we were there, if it turned out that oil was found. I would never do something like that again. But, back then, I thought I could take on everything."

Marny thinks this innate sense of invincibility came, in part, from the major female role model in her early life, her great-grandmother, who "had a farm and did not speak Dutch well, but was the boss of everybody: children, grandchildren, great-grandchildren, employees, the neighbourhood." Marny discovered, quite belatedly, that the very same formidable great-grandmother had never learnt to read nor write.

Similarly, it took Marny a good long while before she discovered there was a gap in her own armour of invincibility. "The first time I realised I couldn't win was when my father fell ill with a brain tumour and passed away. I was thirty-seven years old and I found myself asking, 'How come I didn't succeed? How come it didn't work this time?' Before that, in my work, my family, I never thought I couldn't do everything. So, it came as a shock that there was something in my life outside of my control. And it's not that I was always looking for control, but I always thought I could fix things."

And, in truth, fixing things seems to have been Marny's superpower. She helped to fix national problems: at the ILO and at Staatsolie when privatisation was threatened. She helped to fix corporate problems, like what to do about the company's "institutional task".

And she also fixed personal problems, like that time when a female employee, who had just been hired and was still on one-year probation, came to her in tears.

"I'm pregnant," the girl said.

"Congratulations," Marny replied.

"But I've not made a full year. They will fire me," the girl said.

Marny's reply was, "Don't worry about that. It would be a fantastic case if they dare. It would reach CNN, I promise you," meaning, she was prepared to fight on the girl's side, if necessary, and to take that fight as far as it needed to go to secure her human rights as a woman.

Marny believes women do have a place in oil and gas, and that employers should face facts. "When you hire a woman, especially a young woman, you must understand that we are productive and we are reproductive. Within a few years young women will get pregnant. You have to incorporate that in the system otherwise you will not get the benefit of them."

Although she perceives no need for a "girls' club", Marny agrees that life in the industry would be easier for women, if they are able to relate to other females in similar positions. She says women need someone who can explain about "the do's and don'ts of an offshore rig, planning babies, organising your family, and how to compete with males who don't have to take care of parents or babies."

Marny says she was able to manage those competing demands because she had a supportive community around her, the central force of which was her husband, whom she knew from high school. Her mother, sisters, and mother-in-law were also part of the support caravan. Marny admits, however, that her social life suffered. "I was once very much involved in the Women's Movement, but I quickly discovered that with the demanding job I had, when I left office, I couldn't do anything else but be there for my family. So, I set aside my activism role and also declined things like Rotary and Lions. That was kind of extra baggage."

No doubt, Marny benefitted from having both the seniority and the personality to stake and defend hard boundaries around her personal life. "When my children were small, whenever I left work, I said, 'Don't call me, I will not answer'. When I left the office, I was 1000% about being with my children. Two were competitive swimmers, they had everyday practice, tournaments throughout the Caribbean. The youngest was into music, so she had to be taken to practice. Me and my husband were very much involved in the children's activities."

Motherhood also influenced Marny's professional development choices. She holds an MBA in Oil and Gas Management from Robert Gordon University in Aberdeen. "The reason why I did it was because my youngest was going off to study in the Netherlands and I was afraid that I would fall into a hole."

Although she doesn't think an MBA is necessary for effective leadership, it does come in handy, Marny says, after a person has acquired a certain amount of experience in the industry. "It helps to level the playing field in certain rooms, especially in strategic meetings. You understand things differently, you speak the same language. The MBA was immediately useful in my work."

Her motherhood skills have also influenced the way Marny manages people. She relates the story of a young trainee, a girl, "small and skinny and from a traditional Indian family", who was adamant about wanting to go offshore when the opportunity presented itself. "In Suriname, Indian girls tend to be very protected, so I told her to discuss it with her parents, but she insisted on going. She and another trainee – a young male – would have had to board the ship to go offshore in the middle of the night, and she was so tiny, I was worried."

So, Marny decided to accompany them, with the intent of showing her face to the seasoned offshore workers, who were all male, and signalling that no nonsense was allowed. "I saw these trainees as my children and I wanted to protect them," she says. It turned out, that girl came to the boat, jumped in with no problem and passed the entire period offshore enthusiastically and without an issue. "She did a great job. The boy, however, wanted to come home after a few days."

That one chap aside, Marny found men much easier to manage than women. "Sometimes the girls were a little bit jealous of each other: straight from university, young, and eager to get in front, they always want to prove themselves – especially among each other. Also, men just go; it was much easier to ask them to do something that was not really well-planned, because they had less strings attached to them. Women sometimes want to know everything in advance and find it difficult to go with the flow."

For Marny, the human factor was always the most important part of being a leader. That's another thing about pioneers: they might be passionate and visionary, but they do understand the value of a good team, and they are able to rally people to a cause. "We were a small group, so I felt very responsible for my team. Sometimes I did worry too much, like with that girl, but I always wanted to know my people...their background, their interests, their baggage... because if I don't know that, then I cannot get the 1000% I need from you."

That focus on the human factor is a skill Marny says she learnt from the Americans who would "meet you once, ask a lot of non-business things, and then, next time, would greet you by asking after your family in great detail." Admiring how they made a point of knowing a person's face and what was behind that face, Marny has always tried to understand everyone connected to her work, from the person cleaning the office all the way up to the directors. She also applied that skill to her dealings with outsiders. "If we had a consultant coming in to do a job, I wouldn't worry too much about the consultant, I would focus on the woman coming with him: she has no girlfriends, her mother is not here, she feels lonely because the guy stays in the field the whole day, and she doesn't know the language."

Insider knowledge is, in Marny's opinion, also what helped her overcome "the boys' club". She became one of them, she says, even if it meant going with them to nightclubs in Rio de Janeiro. "I was the only woman there. Everybody was looking at me, but it didn't bother me that much. It gave me a chance to see the men in a different light."

Marny says she never drank with her male colleagues, but she also never shrank from engaging with them and proving that she knew her trade, wasn't afraid to get dirty, and would not tolerate being seen as a sex object. Her advice to females entering the industry is: demonstrate your knowledge and expertise, and do not back down if they try to make you feel inferior as a woman.

"I always have an opinion. And sometimes that is translated as I want to be right. But no, I don't need to be right, I just want a voice. I don't want to be on my death bed and think maybe I should have said something."

She has an interesting perspective on the purposes of corporate meetings. Very often in the company, people did not want to hear bad news or to talk about failures, but rather wanted to focus only on successes. However, in Marny's view, the whole point of meeting is to be honest. "I would always tell my colleagues, this needs to be a safe space for us. If we cannot be truthful to each other, why are we sitting here?"

She admits that, later in her career, she began to work on making her points in a more diplomatic way. "I recognised that you can catch more bees with honey than with vinegar."

In December 2019, after a twenty-year effort, Suriname made its first offshore oil discovery – a significant one – in Block 58, and the formal market announcement was made in January 2020. Marny cites this moment in her country's history as her greatest personal triumph and vindication as a woman working in the oil and gas industry.

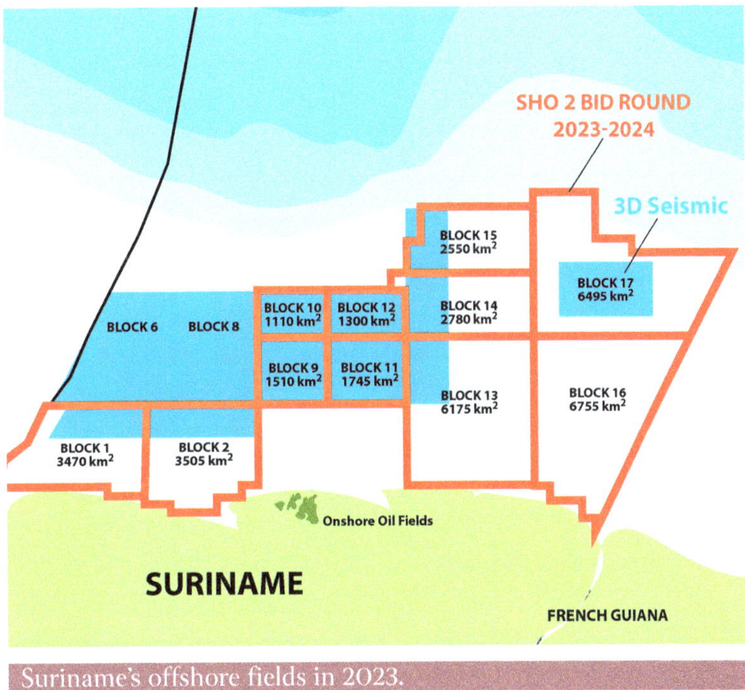

Suriname's offshore fields in 2023.

Of course, she did know that when success happened, the matter would turn political. However, even experienced, savvy, and intelligent Marny had not expected how quickly the very persons who'd been the loudest detractors of the offshore campaign would turn coat and seek to be associated with the success. "Maybe I was naïve. But for so long, no one had been interested, and we had been perceived as a group of people who just travelled a lot and didn't contribute to the overall profit of Staatsolie."

Well, Marny and her team have paid the company back with interest. Although Staatsolie's onshore production is currently the biggest contributor to the government budget, it is estimated to be not even one-tenth of what will come from the first production offshore. So, thanks to the faith, resilience, and doggedness of Marny's team, Suriname's wealth is poised for a huge step change in scale.

Soon after the offshore find, Marny retired from Staatsolie in 2020, at age sixty, feeling satisfied that she'd done her bit, and also hoping that after all her hard work, the new breed of Staatsolie managers will feel a strong sense of responsibility to competently and professionally manage this new phase of the industry. In this regard, Marny predicts Surinamese women will play a key role in the country's petrochemical future "because we are natural collaborators and collaboration is exactly what is needed, going forward."

It is, she believes, the existing oil and gas industry that will fund this planet's energy transition and it is women who will determine the outcome of sustainability initiatives. "Women energise society. That has always been our role. And if you look at traditional societies, it's the women who take care of the energy that the family needs: getting the wood, cooking the food, taking care of animals. Even in modern life, the role of the woman is very visible in energy consumption: it tends to be the woman who decides how to segregate waste in the kitchen, decides where to buy food, and not to buy plastics, etc."

Marny has no patience for people who seek to demonise extractive industries. She believes that, very often, their perspective is hypocritical at best and privileged at worst. The "developed" countries became developed on the back of fossil fuels and now their citizens, sitting on the cushion of their welfare economies, seek to deny poorer countries a similar opportunity for prosperity.

"They are quick to set aside fossil fuels as the bad thing, but then everybody likes to be in an air-conditioned room, everybody likes to travel. They hate the mining industry, but everybody wants a new cell phone and a new laptop. They hate oil and gas, but the industry is much more than what fuel you put in your car; it includes so many petrochemicals we use every day. It includes the fertilisers needed by people in 'developing' countries." Sustainability is crucial, Marny insists, but what we need is to collaborate and move forward to the right mix of energy sources.

Marny says the main takeaway from her life, and the piece of advice she wants to share with everyone – both men and women – is that no one should make a habit of looking back in regret:

"Don't put things in your backpack, you'll suffocate yourself. Something happens, fine. Use your common sense. If we're going through a war, we're not going to walk with a handbook – that's for people who can't think – we just use what we have. Think a little bit: what are the steps you want to take and don't be afraid to take them. If we don't make mistakes, then we don't learn. And when you meet a challenge for the first time, don't think it's the end of the world. There's no dead end, there's always another path. So, don't panic. Get over it, and keep moving forward." ■

CHAPTER 7

VANDANA GANGARAM PANDAY

The Fire is in the Core

VANDANA GANGARAM PANDAY was supposed to follow in her older brother's footsteps. They went to all the same schools and did the same undergraduate degree in Mining Engineering. She even followed him to the Netherlands to do a master's degree, but then Vandana, influenced by the business electives she was taking and by an experience she'd had while doing her undergrad internship, decided to strike out on her own. She returned to Suriname with a Master's in Systems Engineering, Policy Analysis and Management, and since then, she has never worked as an engineer. Nevertheless, she has enjoyed a flourishing twenty-year career at Staatsolie, where, after the retirement of Marny Daal (see previous chapter), she now leads the Staatsolie Hydrocarbon Institute, i.e., the regulatory arm of Suriname's national oil company.

"Why did you break the family mould?" I ask Vandana during our online interview.

"I was doing the graduation project for my bachelor's degree, here in Suriname, with a certain international mining company. I really enjoyed it, but I noticed that people weren't speaking up to the boss when they thought something was wrong, and even when they did, they couldn't get their point across or weren't listened to. I didn't want to become them. So, I decided I should become the boss."

Vandana could not understand their reticence, she says, because she saw the boss simply as a person asking a question and she felt he could be handled

Vandana Gangaram Panday

Vandana and her brother Aroen.

as forthrightly as anyone else. It is only in retrospect she's deduced that their culture of deferral was related to the boss being an expat, white male, seated in Head Office, some distance from the mine site and mine workers; and the company had been undergoing a difficult cost-cutting phase where job retention was of paramount concern to the local employees.

Years later, Vandana had occasion to recall that first job, when, in one of her most poignant career A-ha! moments, she read the notes of her business coach. "She noted in a report on my personal profile, that I don't see hierarchy. I also don't see gender. I see human beings."

Vandana doesn't recall ever being consciously aware of the designations, privileges, or limitations of male and female. She tends to only notice those things after someone else draws them to her attention. "I wouldn't have noticed the gender dynamics at the mine site because I don't record those things in my own system. But later, after going through the whole process of maturity and working in the industry long enough, I must admit gender apparently does matter, but back then I was absolutely not aware. I don't naturally see the difference."

Her gender-blindness does seem inevitable when one considers Vandana's upbringing. She and her brother were raised by a single mother, a schoolteacher who was "busy getting things done" and who had been herself raised by a single mother. Vandana's grandmother brought up nine children alone because her husband passed away early. Both mother and grandmother were very determined, strong women "who just showed up and did what had to be done" without reference to gender roles. So quite early on, Vandana was taught, not by conscious communication but by example, that she had to be independent.

"You have your bicycle, you go to school, you come back, and you make sure you do what you need to do. In school, you get 10 out of 10, and that's not something to celebrate, that's normal. I must admit, that's been who I am, and

I don't get it when people just don't do their job by themselves."

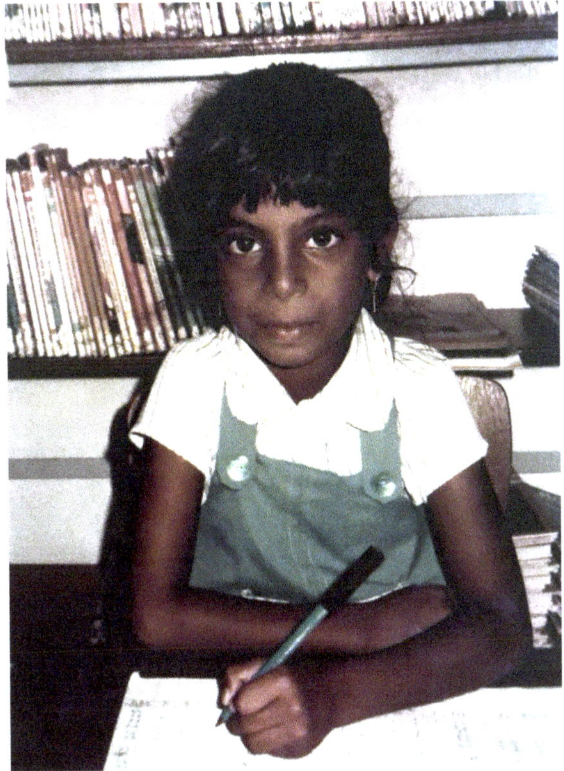
As a schoolgirl.

It's not that Vandana's home was totally devoid of gendered stereotypes. There were some aspects of a traditional Indo-Caribbean tendency to be "a bit more protective" of girls, in terms of where she was allowed to go and what she was allowed to do unchaperoned. However, Vandana says, "I've not been raised not to speak up. It's really difficult, I must say. Sometimes, perhaps I should shut up more, but the way I was raised, we had no time for BS. You do your stuff, you deliver, and you stand your ground if you need to."

"Has what you describe as your 'gender-blindness' and propensity for speaking up been well-received by your male colleagues and bosses?" I ask, and Vandana's response is instructive of her unique personality.

She relates an experience, early in her career, of working for six years on Staatsolie's US$1 billion refinery expansion project, throughout all its phases from pre-feasibility to engineering and construction. It was a male-dominated environment, which she enjoyed very much.

"Every day there was a problem to solve and with males, generally, you can have a good clash and work happily together tomorrow. I'm happy to have an opposing view from time to time. It's part of the excitement, and I truly believe it's instrumental to better decision-making. Also, I am inspired by Hindu scriptures which say, 'It's your duty to fight the righteous battle'. So, I 'fight', decide the way forward, then continue the job toward a common goal, and with males that's easy to do."

Vandana even considers it her "good fortune" to have had, at that time in her career, a particularly clash-friendly working relationship with one of her former bosses. He was a very experienced man, from whom she learnt much, with plenty of room for disagreements. She believes she won his respect and gained significant practice in how to handle conflicts. "It's difficult for the senior male if you hurt his ego, so it creates a clash. However, you cool off, you go back and then sometimes he did change his decision – but then don't talk about it. Otherwise, it's like putting salt in the wound."

After the refinery project, Vandana felt there was a need "to prepare for the future discovery of offshore oil in Suriname, even though it wasn't there yet." An expansion in the industry would require new competencies, she surmised, so she went off to Scotland for a year and obtained a Master's of Law in Mineral and Petroleum Law and Policy. "I loved being a full-time student again, spending lots of time in the library. There's so much more to learn after ten years of work experience."

Upon her return to Suriname, she worked in Petroleum Contracts, and then Corporate Planning & Control. She was then seconded for one year to the national oil company of Norway, where she worked in exploration from a commercial perspective. By the time she returned again to Suriname, in early 2020, the country was celebrating the future she'd been preparing for: its first local offshore oil discovery. She spent the COVID years, 2020–2022, with the Offshore Directorate, as Commercial & Strategy Manager. Although she only took up her current role as head of the Hydrocarbon Institute in February 2023, she has held supervisory roles at Staatsolie since 2017, managing small teams of people – about a dozen at a time.

As far as gender parity goes, at Staatsolie, things are pretty good, Vandana notes. On the current executive board, there are two females and three males. At management level, the ratio is also good, and in the corporate functions there are more women than men. Even in the offshore and regulatory units, there are about forty people – geologists and geoscientists – and more than half are female. "At Staatsolie, in the STEM roles, if they are desk jobs, we also have ladies. However, in the technical fieldwork – operators and mechanics – it is heavily male."

In Suriname, Vandana says, there's a worrying trend at the local university where, except for the engineering degrees, there's a preponderance of females in every field of study. "The males are not studying. So, it presents a different kind of gender problem."

Perhaps as a result of this trend, Vandana is seldom the only woman in the room, at Staatsolie – even when she worked on the refinery project, which was male dominated. The first time she recalls having such an experience was when she went to the Netherlands, as part of the Staatsolie Owners Team, for meetings with their engineering contractor. "That was the first time I did consciously notice that there were forty to fifty people, and I was the only woman. Everyone else, besides my Staatsolie colleagues, was an older white male."

Vandana is careful to avoid generalisation when she talks about the experience of managing males. "In the past, it has been both easy and challenging, based on the individuals."

However, she cites an example where, in a former role, she had experienced some difficulty with a male colleague. "When I try to analyse it, he was more senior in his previous roles, older, and very ambitious. Probably, he thought I was not supposed to become his boss."

She's quick to add that being a woman managing other women also brings its own complicated dynamics. "I've had some difficulties before. I'm not sure if it was because of us both being women. Or was it again the particular person? Sometimes two people just don't really match, you know."

Despite the high female presence at Staatsolie, Vandana has had more male bosses than female. In her experience, female bosses were more "task-focused" like herself, so there wasn't much room for friendship.

Over her career, Vandana has noticed a difference in the way displays of emotion are perceived and labelled, depending on whether the person showing it is male or female. She has received feedback like, "Reduce a bit of the sharpness...You have a sharp edge, you need to cut them out or polish them", in circumstances where she's not sure a man would have been critiqued in the same way, and where his "sharpness" might have been perceived as an asset.

She shares a particular example of standing ground and speaking up to a male colleague. "His boss was in the same meeting and told him, 'Oh don't be intimidated by her.' And I had never thought of myself as intimidating. I thought I was speaking up; I was making a point and if you don't get it, I'll repeat it."

Given her lived experience, Vandana felt vindicated recently, when perusing an international report on this topic of emotion in the workplace. "It showed that statistically, in women's performance reports or appraisal talks, the word emotion gets mentioned I think it was 46% and the males was less than 10%. It's a huge difference. A male will get angry, and it doesn't come up in his appraisal because it's seen as a consequence of his rational behaviour and frustration. But when it's a female, it's mentioned as a performance issue or improvement area."

Notwithstanding the implicit biases she's encountered, Vandana has always felt valued at Staatsolie. "I am valued, so even though sometimes people don't necessarily like me for the bossiness, they do value the skill set. So, I have been supported more often than not."

Staatsolie does not have a formal diversity programme. When it comes to mentoring, in many cases, a person's boss ends up being their informal mentor. Vandana has also had an external coach, a female business psychologist, who has helped her to learn a lot about herself, e.g., her conflict-handling style, her blindness to gender and hierarchy, and how to manage not just peers and downward, but also to manage upward. "I was not playing the game well, because I thought if you're up, if you're a boss, you should be better. That affected unconsciously how I interacted with my superiors."

As part of her daily job of being a boss and leading people, Vandana believes that it's not so much what a leader says but what a leader does that makes the most impact on a team. She is always happy to be a "sounding board" or informal mentor for her direct reports or any Staatsolie employee. "It happens a little bit like a one-off, or it may happen through a few scheduled talks in a row. The other person may view it as me mentoring and coaching, but I just see it as, 'Come and have a chat. And feel free to come whenever.' "

To a female entering the oil and gas industry, Vandana's advice is: Be yourself.

"Looking back, the best part of growing up has been becoming happy with who I am, all of it. Personality is a strong force which, to me, you cannot change in the core of who you are. You must take feedback, but you should put more focus on using your strengths than on changing your weakness – those, you work on them in the margin. The energy comes from within, from the core, and you shouldn't let anyone pour cold water on your fire. Be aware that there are male-female differences, but let it not become a theme in itself. If you perform, it will be seen, and you will get recognised for it. Just enjoy the journey of growing because it's fun! Enjoy even the downsides, have a laugh at yourself."

Having broken the family mould in more ways than one, Vandana feels she owes her mother and grandmother a duty, not only to always perform and deliver, but also to speak up and be herself. "I am in a luxury position. I don't

"Just enjoy the journey of growing because it is fun!"

have to struggle like they did. My grandma was de facto illiterate, she couldn't write her name. And my mum was a schoolteacher who had to really work hard during a period which was not a good time in Suriname. I feel grateful for their sacrifices. So, I'm not taking it lightly, otherwise it's like taking things for granted. They didn't go through all that for me to be a doormat." ■

CHAPTER 8

SHARISTA KISOENSINGH

Waar Zijn De Mannen?

Where Are the Men?

HERE IS A LONG TRADITION of female-only places in literature, mythology, and even in the colonial history of South America. The story[1] goes that on June 3, 1542, Spanish conquistador, Francisco de Orellana, was lost on the Southern continent, leading his motley expedition down a big black river. His men encountered an indigenous tribe who erected giant jaguar totems in homage to their fierce women rulers. Orellana dubbed these females "Amazons" – after the legendary all-female warrior tribe of Greek mythology. Farther downstream, he had the opportunity to meet these women when the expedition was attacked by a dozen of them, who were "very robust, and go naked with their private parts covered, and their bows and arrows in their hands, doing as much fighting as ten Indian men." Orellana eventually made it to the river's mouth, Venezuela, and then went home to Spain. When the expedition's chronicles became famous, so did the myth of the South American warrior women. Eventually, the region was named in their honour: the Amazon – a vast biome spanning eight developing countries, of which Suriname is one.

The Amazonian female utopia (or dystopia) subsequently became a common trope in speculative fiction – including movies and TV shows, e.g., the ever-popular 90s cult hit, Xena: Warrior Princess. And that trope is what came

Sharista Kisoensingh

A geologist in the making.

immediately to mind when Sharista Kisoensingh discussed her group of colleagues at Staatsolie, the national oil company of Suriname. Sharista, thirty-six, who is the Team Lead of Deepwater Exploration Contracted Acreage at Suriname's Offshore Directorate, told me that all six of her direct reports – geologists and geophysicists – are women.

"Where are the men?" I asked, marvelling.

"Last year, I had one man, who was rotated for a year in my team. And then in the past, I also had two men. But it has mostly been women geoscientists," she said.

Sharista herself, who holds a Bachelor's Degree in Geology from the local university in Suriname, a Master's in Sedimentary Geology from Utrecht University in the Netherlands, and an MBA in Management and Leadership, has been at Staatsolie and its subsidiaries for almost fourteen years, transitioning from onshore and near-shore work, to deepwater activities when the Offshore Directorate was set up after the discovery of offshore oil in 2019.

The situation she describes in Suriname's petrochemical industry seems the exact opposite of that captured by Christine L. Williams in her book, *Gas-Lighted, How the Oil and Gas Industry Shortchanges Women Scientists*[2].

According to Williams, in 2007, two women, Burek and Higgs[3], writing for the Geological Society of London published the first scholarly book on women in the geological sciences. Their book describes the historical barriers women have passed through: women were not allowed to conduct fieldwork unless a male relative was present; they were excluded from college-level teaching; and those who managed to secure professional positions were fired from their paid jobs when they married (but were allowed to stay on as volunteers). The book also points out that only twenty women were employed as geoscience professors in the United Kingdom in 2007, down from a peak of twenty-five in 2004, when women held 7% of such positions.

Williams' own study, started in 2012, was commissioned by PROWESS, the Professional Women in Earth Sciences, which is a special interest group of the American Association of Petroleum Geologists. The study followed a cohort of three hundred and sixty North American scientists and engineers over a five-year period, involved in-depth interviews with a subsample of forty-four such people, only to conclude that "women leave the oil and gas industry because they are forced out."

However, the accounts of the Surinamese women in *this* book do not reflect such dire prospects. In fact, they seem to indicate that women like Sharista, who also teaches geology at the local university, are on the ascendancy in Suriname's oil and gas industry. She, and her compatriots, Marny Daal and Vandana Gangaram Panday, have explained the situation as stemming from a preponderance of females at the universities.

"Some departments, like the social sciences, have always had a lot of women. But now, within oil and gas, we see since the last five years or so that women have started dominating. So, you will see it in the young generation, I would say up to maybe thirty-five to forty years old, you have mostly women. But if you go higher up, yes, then you still do see quite some men in the STEM roles."

Dominance within the education system translates into the workforce at Staatsolie. "With regards to Operations, that's in the onshore field, you would see more men. Tradespeople, technicians, those kinds of labour-intensive jobs still tend to be male dominated. But if you look at the higher-educated group like the drilling engineers, the production engineers, the geosciences, we have started to dominate or at least be an equal amount, but definitely not less than men."

Gender parity within the oil and gas workforce has led to similarly favourable ratios within management at Staatsolie. "On average, you do see that everyone

Surmounting gender norms ...

gets an equal opportunity. There are a lot of females in management," Sharista says, although she admits that over her fourteen-year tenure, "three to four years, it was with a female boss. But the other ten years, my boss was always a male."

She's noticed no real difference in the management style of her male-versus-female bosses, or in her own management style toward men and women. "It's not a men-women thing at Staatsolie. There is no gender difference in management."

I enquired whether this could be because men have always been in the minority in her work teams, so there was little opportunity for pooling testosterone to fester into gendered behaviours, or to foster unhealthy competition among the females. Sharista, though non-committal, agreed that might be a possibility. She is adamant that she's never felt discriminated against because of her gender. "I've never felt that I wasn't listened to, or my opinion dismissed because I am a woman. Also, I have a very strong voice. It's not a loud voice, but I'm very confident, especially when it's technical. I can make my case."

The status of women in Suriname's burgeoning oil and gas sector not only defies the culture of the global petrochemical industry as described by

... while cherishing tradition.

Williams, but also the traditional gender norms of the country itself. Sharista herself broached this, saying, "Of course, we are from a country where society's thinking overall – and some individuals still do think that way – is that women have a certain role and men should do all the work. But here in management at Staatsolie, that's not the case anymore. Sometimes I might hear some comment or grumbling from the lower-level staff, but it is not something that is affecting me or us in management because we do know better."

Sharista was even more candid about the evolving norms within her Indo-Caribbean culture. She explained that a traditional East Indian background involves females being more protected and avoiding certain types of work and having very traditional roles. That was – and still is – the case in some areas in Suriname, including the remote majority-Indian district of Nickerie, where she was born and raised.

"It's a rice district. The people work in agriculture. As a girl, you're still very protected and there are a lot of restrictions. During my upbringing, I had those restrictions as well. But the funny thing is, in Nickerie, if you want to study, then that is what you are allowed to do. So, they have restrictions on a lot of other things: don't go out in the night, don't go to discotheques. But there's no university there, so all the girls and boys who want to go to the university, they have to leave home at around seventeen, eighteen years old, and move to Paramaribo. Then one out of two things happen: either they study very hard and they finish, or they get freedom and because they're not used to it, they never finish their study."

Either way, once a girl leaves Nickerie and goes to Paramaribo – and attains qualifications or a less-constrained life – she's unlikely to return to Nickerie's rice paddies, and will opt instead to enter into the urban workforce, which will change everything.

However, Sharista says, education and urbanisation do not cause all traditional expectations to fall away. "The expectations are still there: to get married, to have children, to put family first. Especially because the change happened in our generation. So, our parents and our in-laws, they are still very much from the old, traditional thinking. The difference now is that they can only tell us the preferred way, but girls are not forced like in the past. Not as many arranged marriages as before."

Sharista's marriage, which occurred after she had completed her first master's degree, was not an arranged one, and she now has daughters aged two and twelve.

When I asked how she balances the competing expectations of her work and family, she said she found it difficult to answer, as her case may or may not be typical. "I am lucky, I think, to have a very supportive husband who does a lot more than me, in the house, because of my schedule. He is an engineer with a more flexible work role. My parents also moved to Paramaribo, and they help us. But I do see it, within the organisation, even if it's not East Indian, even other cultures as well, I do see that the women are very worried about how to manage work and the everyday needs of their family."

Although she would encourage her daughters, should they someday, express an interest in the oil and gas industry, Sharista has two concerns in that respect. The first is about the possible obsolescence of current learning. "I wouldn't encourage them to get into the traditional oil and gas careers, given how the world is changing; maybe more renewables. Geology and the geosciences may retain relevance, though, because they are now not only focusing on oil and gas, but also more futuristic aspects, like carbon capture."

Sharista's second concern is about reverse discrimination, i.e., a reactionary response to the apparently imminent female utopia in Suriname's oil and gas industry. "Because there seems to be more women in the university, even more women in Staatsolie, gender may be an issue for my daughters when they enter the workforce. What I sometimes hear now is, 'The next person I'm going to hire is not a woman, because I already have so many women.' But I have daughters. If they apply and they are better, but you don't want a woman, then what about them? So, those things do make me feel angry because it is important for me that the qualifications and the competence is the thing driving hiring decisions, not gender."

Life's a marathon that wants to be run.

Her misgivings are real and tangible. If current education and employment trends continue, could Suriname become the first post-male oil and gas economy in the Caribbean? And is that a thing to be desired? We, in the wider region, should be paying close attention to Suriname and to these questions, because our statistics for gender parity in education are far from encouraging.

Williams states that in the United States, the oil industry elite consists almost exclusively of white men, while in other oil-rich countries, the industry is headed by men drawn from the elite ethnic strata in their societies. She also posits that hiring and firing of women in the oil and gas industry tracks the oil price, and that disproportionately laying off women allows companies to revert to virtually all-male bastions after every economic downturn. However, her observations cannot hold true if there are simply no qualified males available – which is the emerging challenge in several Caribbean territories.

A 2022 UN study[4] found that although the Caribbean has achieved and maintained gender parity in access to education at the primary and lower-secondary education levels, at the upper-secondary education level, boys are lagging behind in terms of net school enrollment. At the tertiary

Planning a future in which daughters would be valued based on their impact and importance, rather than on the value of their gender.

education level, the study confirms what is already public knowledge about the underrepresentation of men. In terms of academic performance, girls are outperforming boys, even in subjects and disciplines that are traditionally thought to be areas in which males excel academically. The education system is either failing boys or not incentivising them enough, the UN study says. Either way, boys are increasingly being left behind in the critical race to attain the quality education essential for decent work and livelihood in a world which is rapidly becoming knowledge-based.

A society of underperforming men helps no one. And, in petrochemical-based Caribbean economies like Trinidad, Suriname, and Guyana, an industry of underperforming or occasional men could be just as damaging as one that openly discriminates against women. Of course, there would be macro-level impacts to any kind of significant gender disparity: widening of the skills gap of the Caribbean workforce, falling labour productivity, stagnation or decline of economic growth. But, before we even get to all that, there's the possibility for certain debilitating effects at the micro-level, if men are left behind in the next wave of digital transformation and energy sustainability.

If the global oil and gas industry remains typified by hegemonic masculinity, i.e., masculinity which puts pressure on men to disparage anything "feminine", including jobs dominated by females, there is a risk that Caribbean oil and gas sectors which become female dominated will suffer the wage stagnation historically encountered by "pink-collar" industries.

As Williams[5] puts it, "What do nursing, teaching, social work, and librarianship all have in common? Most of these jobs were once dominated by men, and when they were, they were highly respected and well paid, but as females began to take these jobs, that quickly changed."

So, Sharista and her all-female geosciences and petrosciences cohort may well have cause for concern – especially since, by her own description, their work is not manual in nature but done mostly at workstations with rare visits to the male environment of the rigs and seismic acquisition vessels. The problem is not men; it is society's view on women and "women's work". There is the perception that, once women enter and dominate a field – particularly an indoor, desk-related field – it is an indication that anyone can do this type of job, even though it may require significant educational investment.

She certainly is right to worry about comments to the effect that, "there are too many women, it's time to hire a man," because these are, in effect, signs of a creeping tokenism – which, according to Williams, always benefits men. Women in male-dominated spaces experience many disadvantages of being tokens, e.g., heightened visibility, which can make them feel responsible for the advancement of their gender in that profession, isolation, and pressure to conform to the prevalent culture. However, when *men* enter female-dominated professions and become "tokens", they generally don't experience these challenges. Existing men in the profession tend to take new men under their wing, which gives them a boost professionally, and they are pushed to the positions that have higher pay.

It is Sharista's hope that by the time her daughters graduate from high school and are ready to enter the world of higher education or the workforce, society would be at a place where professions are valued based on their impact and importance, rather than the perceived value of the gender doing the work.

In the meantime, she offers this advice to young women entering the oil and gas industry: "Believe in yourselves. We women usually underestimate ourselves, but we should not do that. Learn to just give things a try and see if you're good at it or not. Consider my example: I come from an upbringing where the women were not that strong and that powerful. And I have been there, many times, and thought, 'Will I be able to do that?' But even if I was scared, I tried it. I moved from Nickerie and lived on my own in Paramaribo and in the Netherlands. I had moments of being scared, but I found what I was good at, and I am happy." ■

Endnotes

1 Robinson, Alex. April 13, 2021. Culture Trip. "The Story Behind How the Amazon Rainforest Got Its Name". https://theculturetrip.com/south-america/brazil/articles/the-story-behind-how-the-amazon-rainforest-got-its-name

2 Williams, Christine L. *Gaslighted: How the Oil and Gas Industry Shortchanges Women Scientists.* University of California Press, 2021.

3 Burek, C.V.; Higgs, . 2007. "The role of women in the history and development of geology: an introduction." Geological Society, London, *Special Publications* Vol. 281: 1–8. doi: 10.1144/SP281.1

4 Abdulkadri, A.; John-Aloye, S.; Mkrtchyan, I.; Gonzales, C.; Johnson, S.; Floyd, S. 2022. "Addressing gender disparities in education and employment: a necessary step for achieving sustainable development in the Caribbean". *Studies and Perspectives series-ECLAC Subregional Headquarters for the Caribbean,* No. 109 (LC/TS.2022/114, LC/CAR/TS.2022/3), Santiago, Economic Commission for Latin America and the Caribbean (ECLAC)

5 Bielby, D.; Williams, C.L. 1995. "Still a Man's World: Men Who Do Women's Work". *Contemporary Sociology: A Journal of Reviews* 24(6): 809. doi: 10.2307/2076713

"Our greatest strength as women in leadership, is we lead not only with Logic but with Empathy, that makes the difference."

Michelle Des Etages
Chief Executive Officer,
Renaissance Energy Limited

Guyana

CHAPTER 9

GRACE HUTSON

Woman, Know Thyself

as told to
Celeste Mohammed

MY NAME IS GRACE HUTSON. I am Guyanese, twenty-nine-years "young", but have always been an "old soul" and a problem-solver. Without reservation, after four years in oil and gas, I can say that this industry has deepened my innate passion for finding – and crafting – innovative solutions.

Currently, I'm an Account Manager for Landmark Software Services at Halliburton – one of the largest service companies in the global energy sector – assigned to the Caribbean Geo Market, covering Guyana and Suriname. In that role, I support digital transformation strategies for oil companies – both large multinationals and smaller nationals – as well as regulatory agencies. You could say, I "sell" digital solutions to meet client's operational challenges. But I also occupy another role at Halliburton, within the Women Sharing Excellence (WSE) employee resource group (ERG), where I focus on the challenges women have traditionally experienced in the oil and gas industry, and collaborate to recommend solutions.

Grace Hutson

Given these dual roles at Halliburton, I've been blessed with a front-row seat to the Caribbean energy industry's ongoing transformation, which is being driven by both digitalisation and Diversity, Equity, and Inclusion (DE&I). I've had an unconventional career path into the industry, and from this vantage point, I have, perhaps, developed unorthodox views about women's empowerment within the male-dominated sector.

My Vantage Point: Digital Transformation

Digital transformation will, across every industry, redefine work methodologies and reshape business practices. My job involves the identification of opportunities and the execution of strategies to infuse digital technology across all facets of the Exploration and Production (E&P) business. The intention is for my oil and gas clients to experience a marked shift in how they operate, resulting in enhanced efficiency, safety, scalability, and process optimisation – all of which compound to maximise their overall asset value. My job doesn't entail having subordinates in the traditional sense, rather, it is all about collaboration – both externally, as I pay keen attention to clients' needs, and internally as I convey that information back to my team crafting the digital solutions.

My Vantage Point: Diversity, Equity, and Inclusion

More and more, the energy industry is recognising the importance of diversity, equity, inclusion, and meaningful youth engagement. Within Halliburton, Women Sharing Excellence employee resource group provides professional development tools and opportunities to encourage women's contributions, and to help achieve Halliburton's goals for growth, profitability, and leadership. As Chair for the Caribbean section of WSE, I help to collate the experiences of women within my region, as well as to disseminate information and support services. I'm also a board member of the Society of Petroleum Engineers, an international association for energy professionals, where I serve as Secretary of the Georgetown section, as well as the Coordinator for Energy4me, the educational arm.

My Vantage Point: The Proliferation of Non-Technical Roles

There's also another way in which my presence within the oil and gas industry is indicative of the evolution occurring in the energy sector. Typically, one might envision someone within this sector holding an engineering degree, such as petroleum engineering or petrophysics. I am a graduate of the University of Guyana; however, my educational background is an amalgamation of social sciences, business, project management, and a complementary technical focus. I hold a Bachelor of Science in International Relations, an MBA focused on International Business Strategy, and a supplementary Project Management certification. I'm currently pursuing certification in Petroleum Data Management. Collectively, these credentials converge to support my role at Halliburton. Each day, I witness the harmonious interplay of all four facets contributing to my success. While I hold no regrets about not commencing with a STEM-related or STEM-intensive qualification, I do recognise the importance of pursuing industry-recognised credentials.

Know Yourself and Build on Past Experience

In 2019, while working for an international marketing company on a short communications project for oil and gas, I was exposed to Halliburton and was encouraged to apply for a particular job opening. I responded to it, did the interview, but came away with the feeling that the job was not a great fit. It was my awareness of what I'm good at and what I'm hoping to learn more of, that allowed me to confidently decline that job.

I left Guyana for an opportunity overseas, during which, I stumbled upon the present opportunity with Halliburton. It piqued my interest, so I applied, did the interview, and immediately felt the job had the right number of challenges "to keep me on my toes". My pursuit was not solely to find a position that complemented my strengths, but also one that would push me and create opportunities for continuous learning. This job offered that prospect while allowing me to leverage my previous work experience and my two degrees. I joined Halliburton in January 2020 and have remained on the team ever since.

Working in the oil and gas industry might have been new but the business principles/skills I had been acquiring and practicing for years remained applicable; they merely needed to be applied in a new context. I'd started

Sharing insights with other professional Caribbean women at the International Energy Conference and Expo Guyana, 2023.

working at age fifteen through a work-study programme and had been assisting my parents with their business ventures even earlier. By the time this opening at Halliburton arose, I'd already had a three-year tenure as General Manager of a media communications company, as well as a posting with an Irish/French firm specialising in project management for fibre optics. I've always had a passion for work, learning – and, of course, earning – and this lengthy exposure to the business world not only matured me, but gave me discernment and conviction as to what I wanted and didn't want in my career path.

Clarity begets confidence. Perhaps the best way to illustrate what I mean is to share an incident which occurred shortly after I'd secured the Halliburton offer. While attending an oil and gas conference in Guyana (prior to officially starting the role), I visited the Halliburton booth and ended up having an exchange with one of the company's senior executive leaders, who asked me a series of out-of-the-box and hypothetical interview-type questions, culminating with: *Why should we hire you?*

I replied, "Why should I work with you? I am firm and confident in who I am and what I have to offer, and I know that at this time in Guyana, businesses are seeking qualified Guyanese who have a certain amount of experience. I've seen several enticing opportunities, so what sets your organisation apart?"

With apparent surprise, he chuckled and said, "You know what: for that very answer, I am happy we've made the decision to hire you. You're not just looking for a job, but you're really looking to ensure that you can offer to the job just as much as the job can offer to you."

But First, Acquire Experience

In my opinion, introspection and self-awareness are key to knowing whether you can evaluate a job as much as it is evaluating you.

For an individual who is freshly graduated and contemplating a career in the oil and gas industry, I would offer the following advice: even if you're uncertain about your specific career path, consider applying for positions that align with your degree. In the absence of a well-defined sense of purpose or a concrete plan, seize any employment offer that comes your way. Opportunities within the industry are abundant and diverse, spanning both technical and non-technical roles. By accepting a job offer within the industry, you expose yourself to many potential career avenues. What you initially believe to be your ideal role, may, in practice, prove to be a mismatch. You might discover an unexpected passion for something entirely different and have the opportunity to pivot. If you're at a stage where gaining experience is paramount, don't hesitate to accept a job offer.

However, if you possess a clear vision of your identity and purpose, then it's imperative to be discerning about how you allocate your time and energy. Be mindful of accepting a job that doesn't align with your overarching goals, as such a decision may simply waste your time and limit your learning potential.

Use the Skills Developed in Childhood

It's no joke: confidence is everything. In my case, confidence comes from my childhood and the way I was raised. My mum, despite lacking a formal degree or a high school certificate, exuded excellence and assurance in every endeavour, regardless of societal constraints. From an early age, she told me, "There is nothing under the sun that you cannot do if you really want to do it. If you don't achieve it, it's because you never really wanted it and you have yourself to blame. If you really want it, and you pursue it diligently and wisely, you will achieve it."

Dressed for success.

She also reinforced the importance of education. In her words, "For the world you will inherit, anything that you want to achieve in this life, you need to be educated. Education is going to give you the opportunities, the exposure, the ability to earn."

My dad nurtured my confidence as well, but in a different way, and perhaps, with a gentler touch. In fact, it seemed my parents switched their parenting approaches as I entered my preteen years. My mum, driven by the idea of preparing me for a competitive world, took a less mothering, more assertive stance. Whereas my dad adopted a more nurturing approach, consistently reinforcing that I was beautiful, loved, and capable of anything. His influence has been invaluable, shaping my relationships with male figures, including my brothers and male friends.

There were six of us siblings. I am the third, the second sister. My parents taught us, very early on, the importance of being responsible. I was managing finances in my house at the age of fourteen. My mum would give me the money and send me to pay bills, or I might go to the bank with her, but I would be the one to speak to the teller. In other instances, I would be the one to go to the supermarket: she would send me with a taxi and a list. And I would go, they'd pack everything in the box, and I would double-check the list and do my math. I took that job seriously because I wanted to be that person. In a way, I did not always connect with being the playful child with no real responsibilities. I really connected with being the person who stayed on top of things and was always in control. My parents did an amazing job of instilling an assurance that, in any situation, I will always be able to adjust and manage.

My dad was involved in construction, so I supported him, particularly with the preparation of estimates. I have a good understanding of some aspects of construction work. I also worked for my mum's business, helping with purchasing and sales. I wanted to have my own money to manage. I did not like the idea that every time I needed something I had to go to them – despite having stipends. My parents did not believe in just giving out free money, so they said, "You want more money? You're gonna have to work for it, because nothing in life is free."

Many of the skill sets I rely on in my current role were seeded at home: the sense of responsibility and organisational acumen, task-oriented focus, and a determination to see a job through to completion. My adaptability and commitment to learning is what enabled me to develop new skills faster, e.g., my selling skills. Also, an integral aspect of sales is the confidence to communicate effectively, connect with clients, and navigate rejection. Thanks to my parents, I've never been one to give up when met with a "No".

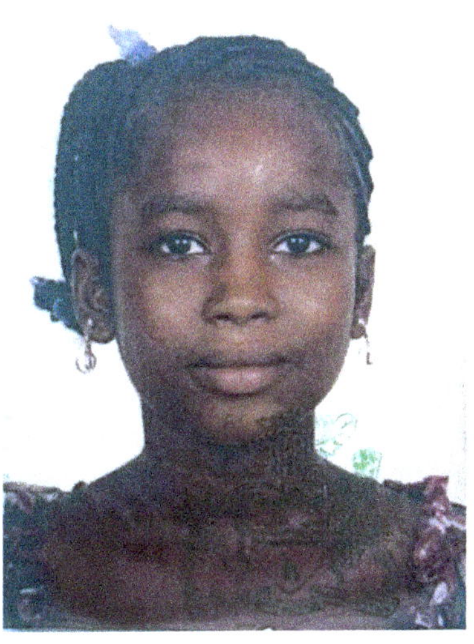

Confidence personified.

Beware: Confidence Can Attract Competitiveness

Women pursuing success, particularly in a male-dominated industry, will at some point confront the issue of female competitiveness. It exists in many organisations. It develops among women who lack self-assurance and confidence in their abilities. Typically, these women don't understand that each of us has unique traits to offer, that my journey is not your journey, that what is for you is not for me. If another woman gets ahead before you, bless her! Consider that maybe your journey was to go somewhere else or maybe you needed to learn something else, maybe your opportunity will come at a different time. If you don't maintain that kind of broad perspective, you will end up viewing every successful woman as your competition.

I've come to understand that encountering a woman in any space doesn't inherently mean that person will be serving the interests of other women. While female representation is undoubtedly essential, being a woman and seeing another woman is not enough. There have been instances where I've witnessed men being more supportive and espousing more positive ideals than some women. For me, the focus should not be about having more females or having less males. Can we get a group of people who are like-minded, equally open-minded, equally encouraging, and supportive? I believe a collective shift in attitudes, ideas, and ways of thinking is what will further gender balance in the industry.

Heal from Toxic Female Relationships: The Need for Grace

As I'm growing older, I am also coming to terms, in a merciful way, with the knowledge that much of the tension and hurt I experienced in female relationships didn't have anything to do with me. With maturity, I've learnt that I need to epitomise my own name: I need to give "grace" to women who can only relate from a place of envy, covetousness, manipulation, or any other unhealthy approach. They haven't pursued their self-awareness journey, and they're grappling with personal challenges. Despite how they treat me, I need to ensure that I remain kind and respectful. I need to forgive, extend grace, and where applicable, space to them.

That's one of the things I try to reinforce with the mentees I have now. I say to them, "If you were ever this person, forgive yourself, and forgive those people who did it to you." I feel it's important to help young girls process their hurtful experiences because you shouldn't have to go through life with an unfavourable outlook on your own gender.

Healthy Female Allyship: Find a "Celia"

Now, as I'm a bit more selective about friendships, I have a few key women in my close group of friends, as well as a few older women from my church, who I can lean on. And thankfully, in my current working environment, I have a colleague named "Celia".

Everybody should have a "Celia", which I define as the person in your organisation who is so supportive, so confident, so committed to ensuring that you don't make the mistakes she did, that she is going to be honest with you, but by the time she's finished, you are ready to take charge; and the only expectation Celia has, is that you do the same thing for somebody else.

My "Celia", Celia Garcia James, has seventeen years' experience in the oil and gas industry. She is a mother of two beautiful kids. She is a wife, an advocate, a volunteer. From the moment I joined the industry, she has been the MOST supportive person I know. She recognised that I am where she once was, and she's been adamant in sharing her experiences so that I can learn from them. She has been one of my sounding boards, my occasional reality check, an active listener, an advocate for gender equality, a cheerleader who celebrates and amplifies women's successes, and she never misses an opportunity to expose me to her network. Whatever she gives, it's with, "Grace, take this and run with it." She has been a great aid to my professional growth and personal healing.

I think if more women can encounter a Celia in their life, the world would be a better place, truly. So, I am on a mission, on every platform I get, to emphasise and highlight how important it is to have supportive connections with other women to whom we can turn for advice, without ever having to question their intention. Without having to worry about ego, envy, jealousy, and competition. Women need to have that and to be that!

With my mentor Celia Garcia James.

Healthy Female Allyship: Be A "Celia"

Within Halliburton, I'm the Mentoring Chair for Women Sharing Excellence, but on a personal level, outside of work, I do have two mentees. I don't know these girls by any connection. I met them randomly at conferences, they reached out with questions, and I've been working with them ever since. I decided to do that because, when I consider my younger years, I realise if I'd had someone who was not too much older to share a certain perspective, I wouldn't have made half of the mistakes that I've made. I wouldn't have questioned: *Who am I? How do I want to show up? Why am I so different? Why am I not liked?*

If I can help another young woman navigate that period of uncertainty from mid-teen to early adulthood, I want to do it.

There's one request I have of my mentees: find another young woman and engage with her in the same way. Let's help each other. Yes, you can learn from your mistakes, but I believe there's even more wisdom if you can learn from the mistakes of others.

Healthy Female Allyship: Accountability and Competence

My perspective on diversity and inclusion has been considered unorthodox. For instance, I believe that in addition to addressing gender biases and systemic discrimination, women should also hold themselves accountable for their actions. Instead of solely pointing fingers at men, we must recognise that in various ways, women can also contribute to unnecessary obstacles and act as gatekeepers. Therefore, when some women call for a "girls' club" in response to the "boys' club", I disagree with this notion. Both concepts are exclusionary, and if our goal is to create a more inclusive environment, forming any club not based on competence and building support will not be a solution.

Healthy Female Allyship: Utilise Employee Resource Groups (ERGs)

As a woman in the oil and gas industry, gender diversity and inclusion are important to me and is one of many things I appreciate about Halliburton. The company's commitment to understanding the obstacles women face as they prepare for leadership roles allows the Women Sharing Excellence (WSE) employee resource group (ERG) to implement initiatives that help women achieve their career goals. WSE focuses on initiatives to attract and retain female employees and champion equal opportunities for women through career mentoring, networking, professional and personal development, and involvement in the local community.

I currently serve as lead for the WSE Caribbean division, which is part of the WSE Latin America chapter. WSE Latin America, with the Caribbean division, collate ideas and experiences from across the region to promote gender diversity and improve working conditions for women. One of the benefits of being part of an ERG is access to leadership. When WSE Latin America notices trends that do not benefit women, the group can escalate the situation, recommend solutions, and collaborate with leadership to affect change.

One of the most effective solutions is the WSE mentorship programme. The programme pairs WSE members with experienced mentors who work with mentees to develop competence in specific areas over a designated period. It elicits the experience and advice of both male and female mentors. Mentorships provide a unique opportunity to form alliances, encourage allyship, and share diverse perspectives. If a male mentor and a female mentee can share and understand each other's perspectives while also learning from one another, both benefit from the mentorship. They learn how to be more strategic and increase allyship.

I show up as my best self.

DE&I Prospects for the Industry

While the industry has made strides in attracting women, there is a concerning trend in reports indicating that women in their thirties often resign or transition to other industries. I know several women who have followed a similar path. Many have assumed roles as mothers or wives or have undergone significant family changes that render their current work structure unsustainable. Some of these women were offered leadership positions, but the lack of flexibility within those roles led them to decline, not because they didn't desire leadership positions but because the positions failed to accommodate their need for work-life balance and a supportive environment. To retain female talent, companies must address these challenges, focusing on work-life balance and creating a supportive working environment. Without solving the retention problem, there won't be enough women to fill leadership positions. The industry has been having this conversation for so long, I think unless we start tying DE&I initiatives to incentives or penalties, nothing will really change.

Some companies have made an effort, e.g., using quotas to encourage female involvement, but I am not in favour of any quota where the sole criterion for the position is the gender of the candidate. Leaders, regardless of gender, should be chosen based on their competence and demonstrated abilities. Placing a woman in a leadership role solely because of her gender is counterproductive and does a disservice to employee morale. A woman in any position needs to know what she's doing.

Being a Woman in the Industry: Take Nothing Personally

I have been in situations where my competence was questioned solely due to my gender. However, a fundamental rule in any business is not to take such challenges personally. There was a time – maybe within my first year in this industry – when I used to prioritise proving a point, but then I realised that it is not my responsibility. What I will do, however, is show up as my best self, I will do my job, and at the right time, those who are open to it will recognise me for who I am.

Being a Woman in the Industry: But Do Learn Personal Lessons

When I started getting involved in the technical side of Halliburton, it was almost as if I'd been thrown off a boat. I had to learn to swim! This was excellent training experience to develop endurance. So, my greatest triumph thus far, while working in oil and gas, has nothing to do with the opportunities I've been able to generate for the company, it has been how I have learnt to adapt to uncomfortable situations or unfavourable conditions and still succeed. Sure, I was fleet-footed before, but this industry has taught me that if you normally take two seconds to adapt, you need to now do it in milliseconds. When a situation comes up, I now instantly think: *What is this supposed to teach me? What am I supposed to learn from this?* I try to get my mind in that mode before I start going off on a negative trajectory.

My advice to any female entering this industry is fourfold:

(1) **Develop your competence.** The conversation surrounding some men's perceptions of women often revolves around stereotypes regarding our capabilities. Prioritise your professional development and expand your knowledge base, whether it's technical expertise or other relevant skills.

(2) **Be proactive and strategic.** Be committed to learning, seeking out new opportunities, showing initiative. Be open and willing to take on new challenges. Being overly cautious can limit your growth; if it appears that you're always going to play it safe, you're going to be stuck in a box.

(3) **Be visible.** Success isn't solely about hard work and dedication; visibility plays a pivotal role. Seek out mentors, coaches, and sponsors. A sponsor is distinct from a mentor or coach. A sponsor not only understands and believes in your potential, but actively advocates for you and promotes your name within the organisation. Cultivate meaningful relationships within and outside your organisation, such as with professional groups like the Society of Petroleum Engineers.

(4) **Practice job enrichment.** As you enter a new role, seize any opportunity to engage in tasks outside your typical role. This will provide valuable experiences that can open doors to new opportunities.

Being a Woman in the Industry: Know Your Personal Limits

It is also crucial to be cognisant of your personal limits. In truth, the greatest obstacle or challenge I've had to overcome so far in the industry has been my pride. There was a time, within my first year, when I wanted to demonstrate that I could go above and beyond always ("Anybody want any help? You want anything? Tell me. Come, come, come, come."). I struggled to say "No". I did not have boundaries. I was not always honest about my experiences, and, because of my pride, I wanted to prove a point. It was an unhealthy thing to do. I took on more than I could handle and almost suffered a burnout.

Now, I'm even more self-aware than I was before. I know now how to work smart and what my limits are. I know I should target my "value tasks" in the mornings and "volume tasks" in my afternoons; anything beyond a certain number of tasks is going to affect me and my quality of work negatively. If my plate is full, I communicate with my managers about re-evaluating priorities as to what gets done first. I've realised there's no need to constantly prove my creativity; sometimes, sticking to effective systems or recommending enhancements is the best approach.

Being a Woman in the Industry: A Wealth of Opportunity for Growth

When I think about my life upon retirement from the industry, I'm not going to measure my success by financial earnings. Instead, my success will be gauged by how I grew as a person, and how I impacted other people. The oil and gas industry has enhanced my ability to do both, and I recognise there's always more to discover or unearth.

This industry is replete with career opportunities – and new ones are burgeoning every day – for men and women who are up to the challenge. There's currently a demand for data scientists, data analysts, cloud computing architects, UX designers, etc. However, as the pace of digital transformation increases, there will be new kinds of roles which are neither the traditional support roles – legal, HR, traditional IT – nor the engineering-specific or geoscience-specific roles. These are exciting times for people who want to enter the industry and help it grow, as they grow.

Women should not be deterred, rather they should come forward, in their numbers, with a willingness to be aware, accountable, and adaptable – to imbue the industry with more competence and with cleaner "energy". ■

"Use your voice effectively. Believe in what you can do. Work hard and respect those around you. Take advice from those with meaningful input and bring someone along with you – be a role model."

Mechelle Smith
General Manager (Ag),
National Petroleum Corporation, Barbados

The Drill-Down

CHAPTER 10

Toward a More Stable Platform

BUCKLE YOUR HARNESS AND BRACE YOURSELF – this chapter is going to be heavy. Heavy on the research, theories, and recommendations concerning women in the oil and gas industry. Light on the personal touch – for that, skip to the next chapter.

If modern commentators agree on one thing, it's that the gendered divisions in today's oil and gas industry have been grandfathered in. The sector sits atop a historical platform of intrepid oilmen and muddy, life-threatening, manual labour. Exaggerated machismo – with all its emphasis on penetrating an Earth which is portrayed as intrinsically feminine – might have been a necessary component of the industry in its first century of existence, but those gendered mindsets and behaviours have become increasingly irrelevant. To use a Trinidadian turn of phrase, "We overs that."

In fact, research suggests that, in this 21st century, the global petrochemical industry is in a state of imbalance, and is therefore underperforming, precisely because it still rests atop this creaking platform of male-centrism.

Globally, the sector faces:

- an ageing workforce of mostly men whose passage out of the industry will precipitate the Great Crew Change;
- an increasingly digitised world which demands new kinds of skills, e.g., advanced analytics, machine learning, and robotics;

- demands for sustainability and energy transition to a more balanced cocktail of fuels – a process which will require innovation; and,
- declining appeal among younger people, attributable in part to its being viewed as a "dirty" industry. A decade ago, oil and gas was the 14th most attractive employer among engineering and IT students; in 2019 it slipped to 35th.[1]

Given the need for new "greenfield" talent and greater innovation, it seems obvious that the oil and gas industry must deepen and diversify its pool of personnel.

One way to achieve these ends – larger talent pool, greater diversity, greater innovation – is to bring in and retain more women. Even if gender parity weren't the socially responsible course of action, oil and gas companies would do well to view it as a strategic priority for purely business reasons.

The Benefits of Onboarding Women

Improved Corporate Profitability:- McKinsey's project, Diversity Matters, found that companies in the top quartile for women leaders are 15% more likely to have above-industry average financial returns.[2] Research from other sources, such as PricewaterhouseCoopers[3], the Peterson Institute for International Economics[4], and Catalyst[5] has shown the same: companies with a significant share of female leaders outperform their peers.

Improved Decision-making:- Diversity of all kinds has been shown to enhance business outcomes – e.g., teams with diverse members process information more carefully and objectively than homogenous teams[6] – and gender diversity, specifically, has a measurably positive effect. Boards of directors with at least three women demonstrate improved communications, greater adherence to codes of conduct, and better criteria for managing strategy and monitoring its implementation. Boards with more women directors are also more apt to focus on gender diversity, employee satisfaction, and corporate social responsibility.[7]

Better Senior Management:- When women are found in senior management, studies show a positive effect on motivation, teamwork, and cooperation. In addition to the financial benefits, women in senior management have a positive effect on managers of both sexes and as strong role models to motivate female middle managers.[8]

Better Safety Records:- Companies with more women demonstrate better safety records than companies with few or no women.[9] Female employees are more likely to follow safety protocols, treat equipment responsibly, and operate safely.[10] Women are more careful drivers, and women-operated equipment requires less maintenance and repair.[11] In general, in addition to acting more safely, women foster a more safety-conscious operating environment.

But Where Are the Women? – Global Perspective

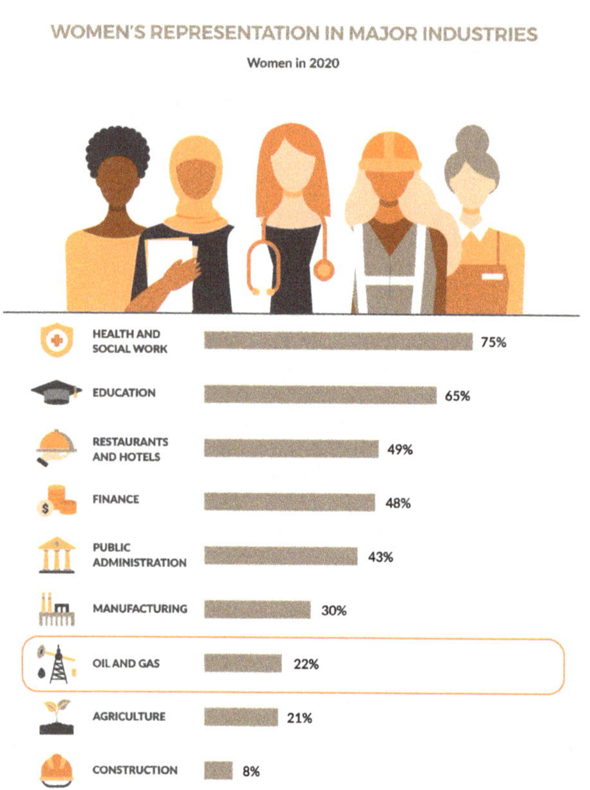

It was only a matter of time before the oil and gas industry would begin to track the research data that gender parity is better for the corporate bottom line. Yet despite the industry's push for Diversity, Equity, and Inclusion (DE&I), which started in the 1990s, companies continue to struggle to attract, retain, and promote women to leadership positions.

A 2021 study by Boston Consulting Group (BCG) found that the percentage of women working in the industry remained unchanged from the level in 2017[12], at 22%. Further, both McKinsey[13] and BCG, after logging the well of female talent within the industry, reached the conclusion that women remain a qualified yet under-tapped resource:

Under-represented in Entry-Level Jobs[14]:- While 44% of male and female STEM graduates and young professionals worldwide indicate an interest in working in the oil and gas industry, women occupy only about one-quarter of the industry's entry-level positions. This results in a slimmer pipeline for women to be promoted from within.

Under-represented in Technical/Operations Roles and Expatriate Positions[15]:- Women continue to be under-represented in these areas – both of which are critical to improving a person's promotion prospects – despite the willingness of nearly two out of three women to accept international assignments and alter their career plans. Similarly, women account for less than one in eight technical and operations employees at all career levels. Women remain concentrated in business and administrative roles.

Under-represented in Senior Roles[16]:- The proportion of women in senior-level decision-making positions is half that of women in mid-level positions. Compared with eighteen other industries, oil and gas was last in female participation at entry level and second to last in the C-suite. When compared with other science, technology, engineering, and mathematics (STEM) industries, it ranked last.[17]

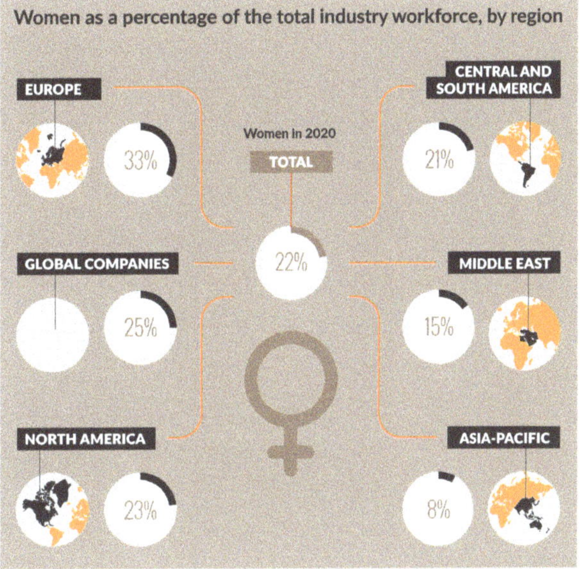

The resulting pipeline of female talent shows not enough women at source, a hollow middle, and a drastically narrowed conduit at the most senior levels.

But Where Are the Women? – Trinidad and Tobago

Something that we must always guard against in the Caribbean, given our unique historical and geo-political circumstances, is wholesale adoption of the results of international studies which may not have included us. However, it is interesting to note how closely Trinidad and Tobago seems to mirror the global statistics discussed above.

In 2017, women in Trinidad and Tobago's energy sector (petroleum and gas sector, including production, refining, and service contractors), accounted for less than 20% of that segment in the workforce.[18] The T&T Energy Chamber has reported that the downstream Contractor[19] labour force in 2021 remained predominantly male, with just 12% being female.

I was unable to locate industry-specific statistics showing the proportion of women leaders in Trinidad's energy sector, but the paucity of women in leadership roles is apparent at the annual Trinidad and Tobago Energy Conference. Anyone attending will notice the "manels", i.e., majority-male panels. The speakers have typically been male, with female speakers accounting for only about 10% on average over the five-year period to 2018. However, in 2019, there were 20% female speakers. At the Clean Energy Conference in 2018, women speakers had 28% participation.[20]

Beyond the sometimes-limited statistics and data, we can look to the lived experience and observations of the women interviewed in this book. These firsthand sources have, in a Caribbean context, intimated exactly what the global research shows: that women are under-represented at every level of the industry; and that the women who do exist are concentrated in non-technical roles. What's more, the accounts of the women in this book suggest that the causes of female under-representation in the Caribbean petrochemical industry match those identified in the global research:

> *Persistent unconscious biases and gender-related challenges combine to either dissuade women from participation in the oil and gas industry or cause them to transition out altogether.*

Why so Few Women?

Unconscious Bias

Unconscious bias has been defined as "unintentional and automatic mental associations based on gender, stemming from traditions, norms, values, culture and/or experience"[21]. It often creates a perception gap between men and women. Given that men hold over four-fifths of decision-making senior and executive-level positions in the oil and gas industry, there continues to be a risk that lack of understanding of the actual barriers women face may negatively impact our prospects for advancement.[22]

For example, unconscious bias against women shows up in the recruitment process – the way job descriptions are worded (e.g., the use of "he" pronouns, "foreman", and other terms that connote traditional notions of masculinity), the tendency for all-male slates of candidates and hiring panels.

Once a woman is hired, bias may also show up in work assignment selection, performance assessments, and promotion decisions. McKinsey found that whereas men are typically evaluated for their future potential, women are typically scrutinised based on their past achievements.[23] Unconscious biases also modify the perceptions of how women and men behave in the same situation. For example, women may be seen as stuck in "analysis paralysis" versus a man in the same scenario who is viewed as a careful thinker.[24]

However, let's not forget that quite apart from the systemic biases of oil and gas companies, in society as a whole, unconscious bias against women exists and operates to constrain our entry into the industry. The erection of these invisible barriers begins in girlhood, and they become embedded in socialisation and learning processes. There are social, cultural, and gender norms, which influence the way girls and boys are brought up. These factors shape their identity, beliefs, behaviour, and choices, including the idea that STEM are "masculine" topics and that female ability in these fields is innately inferior to that of males. This affects the decision of females to enter and stay in a perceived STEM industry[25] such as oil and gas. With the increase in technology and automated operations, however, the traditional argument that women lack the physical skill to do oil and gas work is increasingly irrelevant.

The assumption that STEM-related degrees are necessary for success in the energy industry is often ill-conceived.[26] Although this is arguably true in the fields of engineering, geophysics, geochemistry, metallurgy, and other technical subjects, many people working in oil and gas at all levels do not have a technical background. Non-technical expertise can also be of key importance to management positions. Students and young women may not know this breadth of opportunity exists – I certainly didn't.

Double-Burden?

Women do three times more uncompensated work – housework and caretaking – than men, which hinders many women from reaching their full potential in their professional careers.[27] This is what experts cite as the "double burden," that many women employees are expected to juggle the formal labour sector and unpaid work. After doing their regular job, they are often expected to come home and perform housework or be the primary caretaker of elders or children.

According to the T&T Energy Chamber, the COVID-19 pandemic was a greater shock to the female contractor workforce[28] than the males, "...probably due to the fact that women have had to typically take on the brunt of additional childcare responsibilities while schools remained closed, given the prevalence of continued household division of labour in TT."

Furthermore, working mothers face an additional level of bias, as other colleagues may view them as being less dedicated to their careers than female employees without children.

In truth, for many women in the oil and gas industry, the burden may be well more than double, considering all the instances other than caregiving in which they must work when their male colleagues don't have to. Maybe, as suggested in Chapter 4, we should call it a "quadruple burden". Consider, for example, the work involved in managing the overt sexism of the male-dominated environment.

Conscious Bias: Hostile Sexism (non-physical)

Examples of hostile sexism (some of which were disclosed by the women in this book) include male employees using sexist language or insults, making threatening or aggressive comments based on a woman's gender, harassing or threatening a woman for defying gender norms, treating women as subordinates based on their sex or gender, and punishing them when they "step out of line". In short, language is weaponised against women, and because those attacks may fall short of standard definitions of "sexual harassment" or "discrimination", there is no recourse.

Conscious Bias: Sexual Harassment

There's a tendency to think of sexual harassment in terms of men making unwanted sexual remarks or advances, circulating pornographic images, etc. "Just ignore them," is typically the advice given, but the women affected by harassment do suffer psychological harm.

Furthermore, due to the physical nature of operational work, the hazardous locations and conditions at worksites, sexual harassment may become life-threatening for women working in oil and gas, in ways it would not in other industries.[29] Pranks and gendered behaviour that may seem harmless to the perpetrators, e.g., tampering with tools and communication devices, may put a woman at risk of death or injury by rendering her unable to hear instructions or communicate with co-workers and supervisors, or prohibiting her from completing her work.

Lack of Accommodation: Equipment, Facilities

The clothing, equipment, and personal protective gear used in the extractives industry, historically, have been designed for the physical attributes of men, and not to accommodate women.[30] Machinery, tools, and other equipment are designed for the male physique, and use of these tools and equipment by women exposes them to health and safety hazards and makes them less able to perform their duties efficiently.

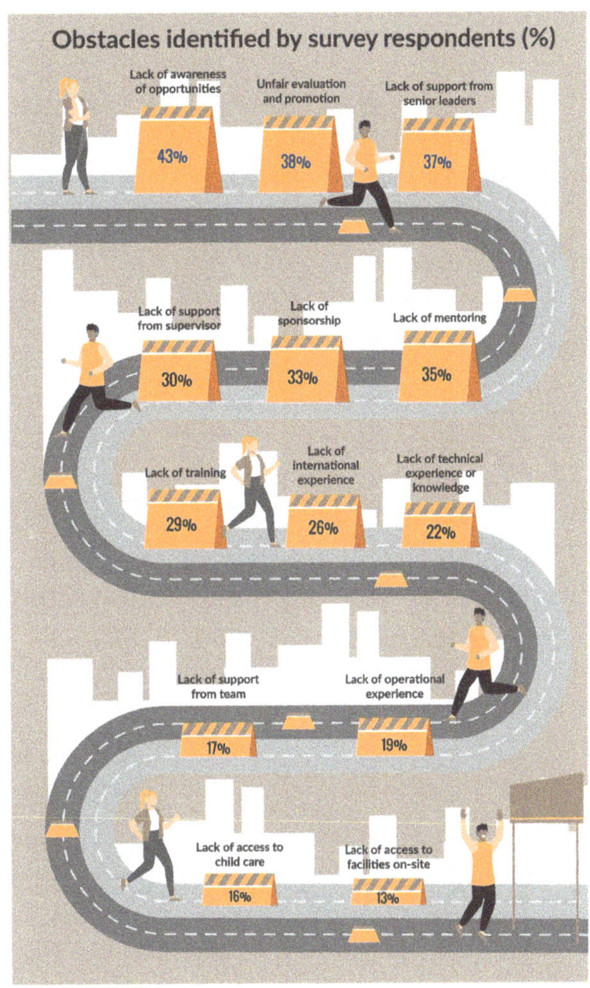

Women in the extractives industry also face a lack of bathroom facilities and change houses or, where they do exist, they are male-designed and without privacy. As mentioned by Satira Maharaj (Chapter 5), uniforms are not constructed to enable women to easily go to the bathroom. Work-around strategies, like limiting water intake during the workday, only pose further harm to their health.

One study has found that the persistent psychological and organisational factors within the industry are "considerably linked" to menstruation disorder and may ultimately affect the sexual and reproductive health of female workers.[31]

The cumulative effect of all the above is a working environment within the oil and gas industry which can so affect a woman's health, well-being[32], and family life that she feels compelled to leave the industry for more amenable – even if less profitable – environments.

No woman wants to exist in a psychological state typified by high levels of stress, anxiety, frustration, isolation, and loneliness – all of which inevitably impact her existing family life and/or her ability to have a future family life.

Who Is Responsible for Change?

Shifting the status quo requires an all-parties approach in the oil and gas industry.

I. COMPANY INITIATIVES

In the local Caribbean context, DE&I policies are still relatively new and are concentrated within the larger multinationals, who tend to have extensive programmes to attract, develop, and retain female talent. Oilfield services and contractors tend to have more volatile margins and may therefore be unable, or consider it unprofitable, to invest in DE&I initiatives. But arguably, there are steps that any employer, regardless of size, can take to attract and retain more women. Strong and visible CEO and executive team commitment to DE&I is perhaps the key prerequisite for change.

Attracting More Women

Companies should think carefully about the images and messaging they convey to the public at large and to women in particular. Companies also need to articulate a positive narrative to graduates and entry-level hires on how the sector is advancing energy transition to a "cleaner" mix. Celebrating women's achievements and highlighting in-company female role models through diversity messaging is crucial. However, these should be authentic and honest representations, as it has been found that where women are candid about both the positive and challenging aspects of their work environment, it resonates with other women who are more likely to believe that there's a supportive female presence within the company.[33]

New recruits are not only looking at people on the ground, but in the boardroom and at senior levels. Research shows a positive correlation between the number of women at the board level and how appealing women find that industry.[34]

Retaining More Women

Foundational DE&I Programmes:- In addition to providing the basic and often mandatory-by-law benefits such as maternity leave, equal pay, etc., companies must ensure that DE&I values are embedded in decision-making at all levels, by: interrogating policies and procedures for unconscious bias; and adopting diversity as one of the considerations for new appointments, policies, and strategies. Senior management must not only be heard to espouse these values but must be seen to uphold them.[35]

Self-imposed Targets:- Companies should consider setting their own gender diversity targets for dissemination to employees. This can serve as a way to raise in-company awareness levels on the benefits of diversity. Such self-imposed regulation must be coupled with meaningful tracking and accountability, maybe even linking management and leadership compensation to DE&I goals.

Flexible Work Arrangements[36]:- If young women can see a way to balance career and family, they may be more likely to want to stay at a company, whereas staff without such flexibility are more inclined to narrow their career goals or look for working conditions that meet their needs elsewhere. Even if such arrangements are available, there's a need to go further and ensure staff may feel comfortable using them. Factors such as stigma, a lack of clarity on how to use them, or supervisor resistance, can deter employees from availing themselves of workplace flexibility.

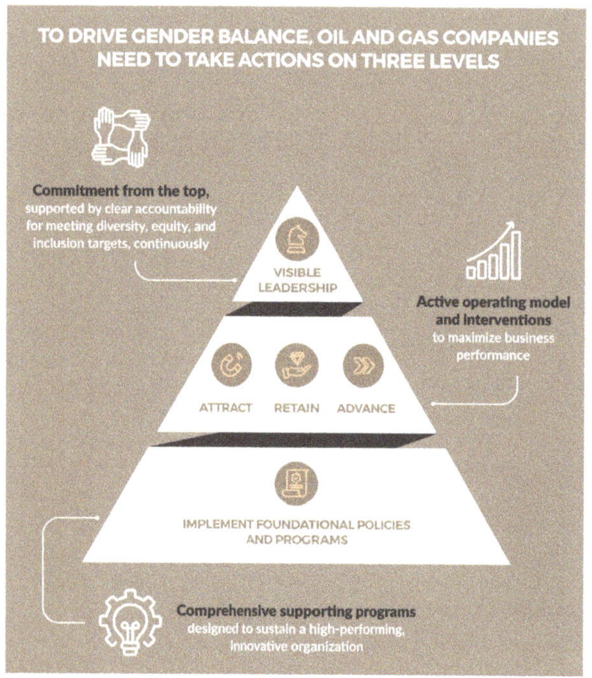

Intake and Career Progression Criteria[37]**:-** As the data and technical needs of the industry evolve, companies should revisit the profiles of their target entry-level candidates: Does every new hire need to have a technical background? Expanding the range of new-hire profiles will create a deeper pool of candidates and possibly attract women – even laterally – from other fields. Furthermore, companies should consider whether it is necessary to perpetuate the long-standing industry practice that experience in technical roles is a prerequisite for senior leadership. Some industry players are revisiting their job requirements and taking the view that a diverse individual with an accounting or commercial background, and strong leadership and strategic skills, may be able to leverage the technical expertise within their team, allowing that individual to succeed in what was previously seen as an exclusively technical role.[38]

And while we're at it, why not re-evaluate the old view that remote assignments are critical for promotion. The industry is undergoing digital transformation and finding more ways to monitor operations remotely. Being physically present in the field may no longer need to be a rite of passage for everyone.

II. INDUSTRY INITIATIVES

Industry-wide change will require industry-wide action.[39] Companies need to come together to collectively develop, implement, and communicate progress. Chambers of Commerce and other industry-wide entities can play a major role in this regard by:

- educating the leaders, i.e., top executives, board members, and shareholders of member-companies, on the value of hiring and promoting more women;

- creating performance indicators that companies of different sizes can adopt;
- encouraging companies to voluntarily adopt and publish scorecards. This would not only create useful transparency about where and how women are advancing, but may also encourage healthy competition among companies to do better;
- compiling a playbook of industry-standard best practices to showcase examples of progress; and,
- hosting and providing women-centred training, sponsorship, mentoring, and advisory opportunities.

One problem, however, with relying only on individual company or industry initiatives, is that some companies may pay lip service to diversity goals without actually implementing the changes necessary to disrupt the systemic sexism built into their organisations. In the context of the oil and gas industry, Christine Williams[40] cheekily calls this "gaslighting". Typical tactics, designed to provide a false sense of achievement and/or attempt to throw critics off the scent, include:

- donating money to organisations or programmes promoting equality that do not interfere with or challenge their normal business operations;
- implementing diversity programmes, such as unconscious bias training, that do not alter the composition of the workforce; and,
- implementing hiring quotas based on a rote check-the-box approach, where companies aim to meet the numerical goal but little else.

III. THE CASE FOR GOVERNMENT INTERVENTION

In our small, multi-ethnic Caribbean societies where the petrochemical industry is (or, in some cases, will soon be) the engine of the entire economy, we can ill afford to rely on the goodwill and conscience of potential gaslighters. Government has an obligation to get involved, where it can be shown that the injection of more women into the petrochemical industry is required by existing law or furthers the public good.

The Public Good: Natural Resource Justice

As multinational IOCs and foreign interests return, in droves, to our shores seeking black gold and its gaseous counterpart, or seeking to establish alternative energy-generation facilities for profit, we, the Caribbean people, would do well to recall the advice Dr Eric Williams gave to post-colonial societies, almost a century ago, in his classic economic thesis, "Capitalism and Slavery"[41].

Dr Williams was writing not just to Trinidad and Tobago, his homeland, but to the entire Caribbean region. To break the yoke of the past, assert ourselves as independent nations, and command the respect we deserve on the world stage, he contended, our governments must:

- redress exploitation by metropolitan interests, be they countries or corporations, by assuming control of how these interests access our natural resources; and,
- admit and address the fact that the relationship between Europe and the Caribbean at the macro-systemic level was mirrored, at the micro-systemic level, by a relationship among peoples based on the ideology of superiority and inferiority.

He noted:

> *The root cause of our difficulties today is as much economic as the difficulties of previous centuries. One group is trying to dominate, everyone is struggling for a greater share of small cake. The solution is very simple unless our community is to tear itself apart – increase the size of the cake and respect the rights of others...*

Dr Williams saw a need for young economies of the Caribbean to refashion our relationship with the countries and multinational corporations at the hub of the international economic system. Secondly, there was a need to increase the size of the economic cake and to ensure that as it grew, the previously marginalised, dispossessed, and degraded workers of the sugar plantation, enjoyed an increasingly larger share. Who can doubt that Caribbean women – be they daughters of former slaves, indentured workers, or freed women – stand among this previously disenfranchised group and still today face ideologies that regard us as inferior in certain economic spaces.

It is important that governments address women's needs specifically, and ensure that we aren't merely relegated to being victims of extractive economic activity or beneficiaries of some amorphous economic gains. Female citizens deservc the opportunity to be actual contributors to the energy industry, on equal footing with male citizens.

There is something to be said for what Bradshaw et al. call "supernormal patriarchal relations" which must be addressed in order to "promote gender equality and natural resource justice, as part of an agenda to redistribute wealth gains from natural resource extraction."[42] Such an approach would, at minimum, require revising the laws and policies that govern the oil and gas industry in the Caribbean (as discussed later in this Chapter).

The Public Good: Sustainable Development

As noted in Chapter 1 of this book, Dr Williams viewed education as one of the crucial means by which the "size of the cake" could be expanded to include marginalised groups in economic participation. When, as Prime Minister, he told the children of Trinidad and Tobago that the future was in their schoolbags, he was speaking to girls as well as boys. Recent data confirm that girls have answered the call, at all levels, including tertiary education and Technical and Vocational Education and Training (TVET). This pattern of female dominance in education can be seen throughout the Caribbean region.

Women and girls are the primary consumers of educational services in the Caribbean. This was the finding of a recent regional study[43] conducted by the UN, which looked at gender disparities in education and employment during the period 2016–2020. At secondary school, although there were fewer girls enrolled in the subjects leading to careers in science, technology, and engineering, girls outperformed boys in Additional Mathematics, Agricultural Science, Integrated Science, Physics, Information Technology, and Technical Drawing. Only in Chemistry did boys outperform girls. The point is: not only are girls studying "hard science" subjects, but they are performing much better than boys in them.

At tertiary level, student enrollment data at UWI indicates that, across the Cave Hill, Mona, and St Augustine campuses of the university, in 1975, there were three and a half men for every woman enrolled. Within only five years, in 1980, gender parity had been achieved: one man for every woman. And by 2015, there was only half a man for every woman.[44] In 2019/20, for every one hundred female students enrolled at UWI, there were fifty-four male students.[45] Females outnumber males in enrollment for most faculties.

Enrollment does not always translate to graduation, so it is useful to consider those figures as well. Females are outnumbering males in getting first class degrees in engineering and medicine. Even in science and technology fields where men outperformed women in the First-Class Honours category, women still outperformed men in the Second-Class Upper Division category.

The educational participation of young women and girls in the Caribbean has not translated to economic opportunities for them.

Yet, the labour market in the Caribbean still favours men. The educational participation of young women and girls in the Caribbean has not translated to economic opportunities for them. This suggests that existing government policies may not be enough to achieve gender parity.

Getting oil and gas industry-specific, we can see this general imbalance is replicated. In the 2019/2020 UWI data, for science and technology subjects the majority of enrollment was by women, 55.5%. Only in engineering do men still dominate, with women representing 34.5% of enrollment. But if past trends continue, this last bastion of male dominance will also succumb. In Trinidad and Tobago, while women's enrollment in undergraduate engineering programmes is still lower than men's, there has been an uptick in women's participation at the post-graduate level, with women representing 43% of students in taught master's programmes, 39% in MPhil programmes and 51% in doctoral programmes (UWI, 2017–2018 Academic Year). Given that women's expertise in technical fields is often challenged in the male-dominated industry sector, these figures indicate that women are seeking advanced degrees in engineering at a higher rate than men do in order to gain a competitive edge.[46]

Yet, as we have noted previously, in the petroleum and gas sector (including production, refining, and service contractors) of Trinidad and Tobago, women account for only around 20% of the workforce. The lower representation of women in this sector is likely also reflected in the composition of leadership and decision-making. As such, there may be less input by women in decisions regarding policy and planning that are related to the future of the industry.

Population data[47] shows that, in 2022, females represented 50.65% of the population of Caribbean Small Island States (in Trinidad and Tobago the amount was 50.68%). Consider too, that by 2050, females are expected

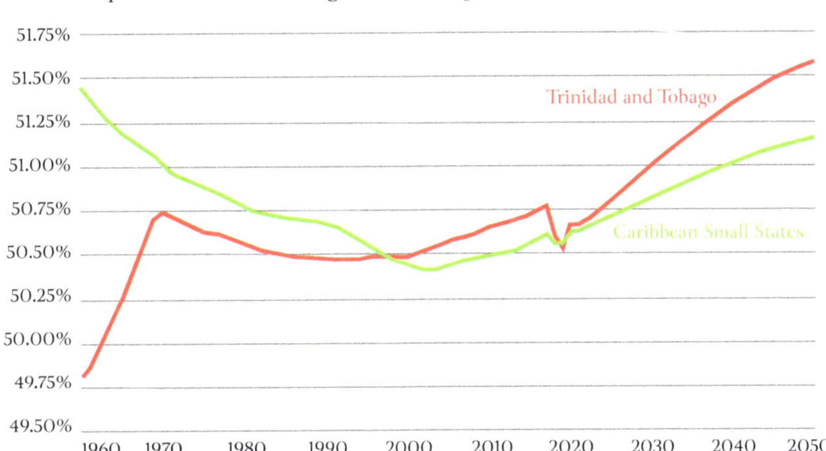

to represent 51.15% of the population of Caribbean Small Island States. As one commentator put it, "...if you're excluding half the population from your recruitment process, you're simply not hiring from the best talent pool available."[48]

If women and girls are not provided equal opportunity for employment in the Caribbean energy industry despite being more qualified and representing a larger percentage of the population and potential workforce, it does not augur well for sustainability as expressed in the United Nations 2030 Agenda for Sustainable Development.

Simply put, Caribbean governments cannot say they are serious about sustainability unless they adopt policies and legislation (and/or enforce those existing) to urgently address the gender imbalances in the human capital development of the region. Such legislation and enabling policies are needed to ensure that women and girls are given equal and fair opportunity to use their skills and attain their economic, political, and decision-making autonomy, which is critical in stemming the rising trend of gender-based violence in the region on one hand, and in innovatively contributing to fill the work force skills gap that continues to exist in the region.[49]

Legal Obligations

Apart from the 2030 UN Sustainable Development Agenda, under both international law[50] and the domestic laws[51] of their respective States, Caribbean governments have legal obligations to guarantee women the same employment rights, opportunities, choices, and benefits as men. They have an obligation not only to affirm the existence of a woman's right to work, but to ensure full and effective enjoyment by women of that right on a basis equal to that of men. Fulfilling these obligations means governments must go further than just "equal pay for equal work", they must guarantee that conditions of work for women are not inferior to those enjoyed by men. This includes:

- the right to safe and healthy working conditions;
- equal opportunity to be promoted based on competence and seniority; and,
- equal opportunity to obtain the education, training, and mentoring necessary to achieve employment and promotion.

It would seem obvious then, that governments should engage in a gender-sensitive review of their existing laws and regulations, with a view to:

- *abolishing* any that restrict the types of work in which women can engage or which permit gender-based workplace discrimination or harassment;
- *operationalising* any legal provisions which do exist for the protection of women's right to work, but have not been typically enforced; and,
- *passing new laws* to that effect.

In Trinidad and Tobago, the most direct and overt forms of discrimination against women have been largely eliminated from the law within the last three decades. However, there is still a need for specific legislation to combat sexual harassment in the workplace, to provide universal maternity benefits, to provide paternity protection (so families have an option other than the woman as the only one staying home), to amend the Industrial Relations Act to ensure inclusion of domestic workers.

While laws are a place to begin, achieving gender equality also depends on the cultural attitudes and norms about women's role in society. There is a direct correlation between gender equality in society and gender equality in work. Thankfully, Trinidad has a whole National Policy on Gender and Development filled with great legislative and practical measures which, if implemented, would take us a long way toward addressing the challenges faced by women in the workplace.

IV. OIL & GAS-SPECIFIC LEGISLATIVE INTERVENTIONS

Local Content Policies (LCPs) have been defined as "a set of policies commonly employed in the petroleum sector to increase the utilisation of national human and material resources, and to domicile in-country oil and gas-related economic activity that was previously located abroad".[52] In short, LCPs are written rules which compel or strongly encourage multinational IOCs to hire and engage local companies, use locally sourced materials, or otherwise involve "local content". They can serve as a powerful tool for governments.

All three territories highlighted in this book have been active in the establishment of LCPs for their respective oil and gas industries.[53] None of their policy frameworks or legislation mention women. This is not surprising, as local content and women's economic empowerment are often viewed as separate areas of public policy.

Sax and Tubb write about the "buzz" and expectations generated whenever there is a new petroleum find.[54] However, this "buzz" may drone out voices calling for a gender-sensitive approach. As one commentator[55] put it, "gender dynamics are a lesser concern than people's immediate interest of seeing tangible benefits from the resource."

However, if it is accepted that Caribbean governments must pursue a holistic approach to gender equality in order to maximise the benefit from petro-development, then LCPs can be used to provide the impetus necessary for more women to be included in the industry:

- directly, by the IOCs themselves employing more women;
- indirectly, by the IOCs tendering rules, etc., preferring suppliers and subcontractors who employ women; and,
- through the participation of women-led micro, small- and medium-sized enterprises (MSMEs) in the petroleum value chain as suppliers and subcontractors.

Such LCPs may take the form of (1) mandatory quota requirements, (2) targets, or (3) tax and financial incentives written into legislation, regulations, and policies. Of course, a prerequisite for all these measures is a pool of competent females with skills geared toward the industry's requirements, and in the case of MSMEs, with the capital[56] necessary to start their own downstream businesses. From the education data shown above, we know the pool of female talent exists and is growing. Capital may be a problem but, surely, governments have it within their power to provide or to incentivise the provision of such financing to female-owned businesses in the downstream contractor supply or services sector.

Internationally, there is some precedent for the use of government's legislative power to address gender imbalance in extractive industries. In Tanzania, the Mining Act (2010) requires that at least one third of members of the Mining Advisory Board are women.[57] Similarly, the 2002 South African Mining Charter included a 10% quota for women in the mine workforce by 2009. Commentators conclude, "The policy has proven successful, with South Africa now boasting some of the highest rates of female participation in mining in the world."[58]

Global precedent combined with the already-existing local content rules in the Caribbean, means that the foundation has already been laid. It's just a question of whether our governments have the political will to commission a wider and more gender-inclusive platform for the petro-development of our small countries. Maybe being small in size, like Trinidad and Tobago, offers the opportunity to be agile and to overachieve; maybe being much-sought-after virgin territory, like Guyana and Suriname, allows for foundational rules to be laid with little resistance; maybe, working altogether, the region can even lead the world in female participation in the energy industry. For the sake of all the little Caribbean girls trudging into school assembly every morning to proudly belt out their national anthem, I hope our leaders will try. ■

Endnotes

1 Yanosek, K.; Abramson, D.; Ahmad, S. 2019. McKinsey & Company. "How women can help fill the oil and gas industry's talent gap". https://www.mckinsey.com/industries/oil-and-gas/our-insights/how-women-can-help-fill-the-oil-and-gas-industrys-talent-gap

2 Hunt, D.V.; Layton, D.; & Prince, S. 2015. McKinsey & Company. "Why diversity matters". https://www.mckinsey.com/capabilities/people-and-organizational-performance/our-insights/why-diversity-matters

3 Women in Mining (UK) and PricewaterhouseCoopers. 2014. "Mining for talent 2014: A review of women on boards in the mining industry". https://womeninmining.com/wp-content/uploads/2014/03/Mining-for-Talent-2014-research-report.pdf

4 Noland, M.; Moran, T.; Kotschwar, B. 2016. Peterson Institute for International Economics. "Is Gender Diversity Profitable? Evidence from a Global Survey". Working Papers 16–3. https://www.piie.com/publications/working-papers/gender-diversity-profitable-evidence-global-survey

5 Von Lonski, U.; Syth, A.; Trench, S.; Goydan, P.; Riemer, P.; Fjaeran, T.; Miras, P.; Merchant, W.; Gauthier-Watson, C. 2021. A collaboration between the World Petroleum Council and Boston Consulting Group. "Untapped Reserves 2.0: Driving Gender Balance in Oil and Gas". https://www.bcg.com/publications/2021/gender-diversity-in-oil-gas-industry

6 Rock, D.; Grant, H. 2016. Harvard Business Review. "Why Diverse Teams Are Smarter". https://hbr.org/2016/11/why-diverse-teams-are-smarter

7 Brown, David. 2002. "Women on Boards: Not Just the Right Thing ... But the "bright Thing". Vol. 341, Issue 2 of Report (Conference Board of Canada).

8 International Finance Corporation. 2016. "SheWorks: Putting Gender-Smart Commitments into Practice". https://www.ifc.org/content/dam/ifc/doc/mgrt/sheworks-knowledge-report-executive-summary-bm.pdf

9 Marshall, Lisa. "Are Women the Mining Industry's Most Underdeveloped Resource". *Mines Magazine*. http://minesmagazine.com/8749

10 International Finance Corporation. 2013. "Investing in Women's Employment: Good for Business, Good for Development". https://ppp.worldbank.org/public-private-partnership/sites/ppp.worldbank.org/files/documents/Global_InvestinginWomensEmployment.pdf

11 Women in Mining (Canada). 2016. "Welcoming to Women: An Action Plan for Canada's Mining Employers". https://internationalwim.org/wp-content/uploads/2020/07/WIM-NAP-book-full.pdf

12 Note 5, *supra*.

13 Note 1, *supra*.

14 Note 5, *supra*.

15 Note 5, *supra*.

16 Note 5, *supra*.

17 Note 1, *supra*.

18 Central Statistical Office of Trinidad and Tobago. https://cso.gov.tt/

19 The Chamber defines a contractor employee as an employee of a contractor or service company entering any of the downstream gas industry facilities, including the LNG facility, petrochemical plants, other heavy industrial plants, and power-generation facilities. See: The Energy Chamber of Trinidad & Tobago. March 8, 2022. "Downstream contractor labour force - only 12% female". https://energynow.tt/blog/downstream-contractor-labour-force-at-12-female

20 The Energy Chamber of Trinidad & Tobago. February 11, 2019. "Harnessing all the talent from the energy sector". https://energynow.tt/blog/harnessing-all-the-talent-from-the-energy-sector

21 The ILO ACT/EMP Research note. August 31, 2017. https://www.ilo.org/actemp/publications/WCMS_601276/lang--en/index.htm

22 Note 5, *supra.*

23 Desvaux, G.; Devillard, S.; Labaye, E.; Sancier-Sultan, S.; Kossoff, C.; de Zelicourt, A. 2017. McKinsey & Company. "Women Matter: Time to accelerate: Ten years of insight into gender diversity". https://www.mckinsey.com/~/media/mckinsey/featured%20insights/women%20matter/women%20matter%20ten%20years%20of%20insights%20on%20the%20importance%20of%20gender%20diversity/women-matter-time-to-accelerate-ten-years-of-insights-into-gender-diversity.pdf

24 Cecchi-Dimeglio, Paola. 2017. Harvard Business Review. "How Gender Bias Corrupts Performance Reviews, and What to Do About It". https://hbr.org/2017/04/how-gender-bias-corrupts-performance-reviews-and-what-to-do-about-it

25 Munoz-Boudet, Ana Maria. 2017. World Economic Forum. "STEM fields still have a gender imbalance. Here's what we can do about it". https://www.weforum.org/agenda/2017/03/women-are-still-under-represented-in-science-maths-and-engineering-heres-what-we-can-do

26 As the careers of Giselle Thompson (Chapter 3), Deborah Benjamin (Chapter 4), and Grace Hutson (Chapter 9) demonstrate herein.

27 Note 23, *supra.*

28 Which, in 2020, accounted for 1,000 female workers, a drastic reduction – 50% – from the pre-pandemic period of 2019, when the female workforce was 14% or 2,000 female workers.

29 Chira, Susan. *The New York Times.* December 29, 2017. "We Asked Women in Blue-Collar Workplaces About Harassment. Here Are Their Stories". https://www.nytimes.com/2017/12/29/us/blue-collar-women-harassment.html

30 Park, R.; Metzger, B.; Foreman, L. 2019. The Advocates for Human Rights. "Promoting Gender Diversity and Inclusion in the Oil, Gas and Mining Extractive Industries: A Women's Human Rights Report". https://www.theadvocatesforhumanrights.org/Res/promoting_gender_diversity_and_inclusion_in_the_oil_gas_and_mining_extractive_industries.pdf

31 Note 32, *infra.*

32 Razafimahefa, R.H.; Pardosi J.F.; Sav A. 2022. "Occupational Factors Affecting Women Workers' Sexual and Reproductive Health Outcomes in Oil, Gas, and Mining Industry: A Scoping Review". *Public Health Reviews* Vol. 43. doi: 10.3389/phrs.2022.1604653

33 Note 30, *supra.*

34 Note 30, *supra.*

35 A key finding that emerged from the BCG 2017 survey was that if the CEO considered gender balance important, most male employees in the company would, too. See: Note 5, *supra.*

36 Note 30, *supra.*

37 Note 30, *supra.* Note 5, *supra.*

38 Note 5, *supra.*

39 Note 1, *supra.*

40 Williams, Christine L. *Gaslighted: How the Oil and Gas Industry Shortchanges Women Scientists.* University of California Press, 2021.

41 For a good overview, read: Henry, Ralph M. 1997. "Eric Williams and the Reversal of the Unequal Legacy of 'Capitalism and Slavery' ". *Callaloo* 20(4): 829–48. http://www.jstor.org/stable/3299411

42 Bradshaw, S.; Linneker, B.; Overton, L. 2017. "Extractive industries as sites of supernormal profits and supernormal patriarchy?" *Gender & Development* 25(3): 439–454. doi: 10.1080/13552074.2017.1379780

43 Abdulkadri, A.; John-Aloye, S.; Mkrtchyan, I.; Gonzales, C.; Johnson, S.; Floyd, S. 2022. "Addressing gender disparities in education and employment: a necessary step for achieving sustainable development in the Caribbean". *Studies and Perspectives series-ECLAC Subregional Headquarters for the Caribbean*, No. 109 (LC/TS.2022/114, LC/CAR/TS.2022/3), Santiago, Economic Commission for Latin America and the Caribbean (ECLAC)

44 Ruprah, Inder Jit. October 12, 2016. "Move over Engineers, the Girls are coming!" Inter-American Development Bank. https://blogs.iadb.org/caribbean-dev-trends/en/move-over-engineers-the-girls-are-coming/

45 Note 43, *supra*.

46 United Nations Development Programme. February 19, 2021. "Trinidad and Tobago Gender Analysis". https://climatepromise.undp.org/research-and-reports/trinidad-and-tobago-gender-analysis

47 World Bank Open Data. https://data.worldbank.org/

48 Note 30, *supra*.

49 Note 43, *supra*.

50 E.g., Convention on the Elimination of All Forms of Discrimination Against Women (CEDAW); International Covenant on Economic, Social and Cultural Rights (ICESCR); and the International Labour Organisation (ILO) Conventions.

51 E.g., Constitution of the Republic of Trinidad and Tobago Chapter 1:01; Equal Opportunity Act of Trinidad and Tobago Chapter 22:03.

52 Ovadia, Jesse Salah. 2014. "Local content and natural resource governance: The cases of Angola and Nigeria". *The Extractive Industries and Society* 1(2): 137–146. https://doi.org/10.1016/j.exis.2014.08.002

53 Local Content & Local Participation Policy & Framework For The Republic Of Trinidad And Tobago Energy Sector, October 2004; Local Content Policy Framework, Suriname Trade And Industry Association (VSB) & Manufacturers Association In Suriname (ASFA) & The Alliance; Local Content Act 2021, Guyana.

54 Sax, M.; Tubb, D. 2021. "The buzz phase of resource extraction: Liquefied natural gas in Kitimat, British Columbia". *The Extractive Industries and Society* 8(3). https://doi.org/10.1016/j.exis.2021.100938

55 Wyndham, V.; Lange, S. 2019. "Making Sense of CSOs (in)action in Tanzania's petroleum sector: where is gender?" eds In: Fjeldstad, O.; Mmari, D.; Dupuy, K. (Eds.). *Governing Petroleum Resources: Prospects and Challenges for Tanzania*. Chr. Michelsen Institute & Dar es Salaam: REPOA, Bergen, pp. 128–131.

56 Perks, R.; Schulz, K. 2020. "Gender in oil, gas and mining: An overview of the global state-of-play". *The Extractive Industries and Society* 7(2): 380–388. https://doi.org/10.1016/j.exis.2020.04.010

57 Ovadia, Jesse Salah. 2022. "Addressing gender inequality through employment and procurement: Local content in Tanzania's emerging gas industry". *The Extractive Industries and Society* Vol. 9. .https://doi.org/10.1016/j.exis.2021.101028

58 Note 56, *supra*.

"Stay true to YOU! Keep relevant and abreast of all global industry changes by reading, reading, reading! Attend as many industry-related forums and always be open to listen and learn."
Natasha Fournillier,
General Manager, Chag Terms (Trinidad) Ltd.

CHAPTER 11

Extracting the Lessons

WHEN I WAS A CHILD – this is in prehistoric times before malls, cable TV, and video games – there wasn't much to do in Trinidad by way of recreation. My family would sometimes go for aimless Sunday afternoon drives along the Southern Main Road. Even with my eyes shut, I would know we were getting close to the Pointe-à-Pierre refinery, by the pungent smell which made us kids hold our noses and say, "Oh geed! Oh geed! Why this place always smelling so?" But the odour didn't seem to bother my parents. They explained, with patience and a good dose of condescension, "That's just fumes." Like, just accept it. That's how it is, that's the burden which accompanies the benefit: oil stinks.

Now, as a grown woman, I've come to think it's the same with unconscious bias and sexism. They exist everywhere, like invisible fumes just floating in the atmosphere, but we only notice when enough collects in one location – e.g., the oil and gas industry – for us to call it out and say, "This place stinks!" But even then, when the stench is acknowledged by everyone, not everyone will think it can be changed. Many will view it as simply the price a woman must pay for drawing her share of oil wealth. Trinidadians, in particular, have an inside joke about our legendary sense of inertia, or as it's locally dubbed, our we-like-it-so mentality.

But things *can* change, if enough of us believe in that change.

Having discussed, in the previous chapter, the challenges and barriers facing women in the oil and gas industry, as well as the interventions necessary at the company, industry, and governmental levels, I want to focus here on what we as women can do, to help effect the change we'd like to see, i.e., gender parity in the leadership of this industry. After all, if women's empowerment means anything, it must mean not waiting for someone else – especially a cloistered fraternity of oilmen – to gush forth in white tankers to save us. As Grace Hutson and Deborah Benjamin so eloquently expressed in their respective chapters, we should be prepping to save ourselves.

Our interdependency means, though, that a good way to empower myself (or my daughter, for whose future I'm fighting even more fiercely than my own) is to seek to learn from other women's experiences. I interviewed and listened to the women in this book trying to glean what, if any, traits or characteristics they shared that might explain their staying power in a hostile environment. I hoped I could extract from their narratives some kind of secret sauce for female success.

On the other hand, I was a tad wary. Although their stories had the potential to offer helpful insights about coping and even thriving in male-dominated environments, the women themselves might not be able to accurately identify why they succeeded while others failed. An employee is not always privy to the decisions that affect her retention and promotion – those decisions often involve multiple "bosses" huddled behind closed doors. Also, human nature is such that it can be tempting to explain one's own success in self-serving terms.

Nevertheless, I persisted. I sent all the women the same list of questions, I spoke to each separately, and as of the date of writing, none has seen the others' responses. And yet, the attributes displayed in their career narratives share unmistakable similarities. Running through their disparate histories and personalities seems to be a cord braided from the following strands.

They Are Continuous Learners

Formal schooling is something most of us were compelled to do, up to a certain point. Then we felt impelled to learn whatever was necessary to acquire the qualifications pertinent to our career of choice. But fewer of us choose to engage in continuous learning – as in all the time, every day, every chance we get. Lifelong learners never settle for what they know, but rather, continuously seek to improve and build upon their current knowledge, by discovering what they don't know.

A key lesson from Arlene, Deborah, Marny, and Giselle is that the most crucial piece of information a leader needs to know, is the fact that she doesn't know everything. That makes total sense. It's simply not possible for a leader to always know the answer, or even to be the most informed within a given group of colleagues. A leader who is a lifelong learner will be curious, will solicit new information and ideas, will listen actively, will gather information broadly, and will not allow perceived wisdom or prior belief to constrain her thinking.

This requires humility.

Recent psychological studies have indeed shown that "intellectual humility" is needed for effective learning.[1] Low intellectual humility is associated with the tendency to feel threatened by what one doesn't know[2], and to defend one's ego in the face of ignorance, errors, disagreements, and other signs of one's intellectual shortcomings[3]. In contrast, people high in intellectual humility enjoy finding out new information that challenges what they already think is true. In other words, lifelong learners are likely to make better decisions and therefore to be more successful leaders, as they spend less time relying on and defending what they already know, and more time acquiring new concepts and ideas.

This view was echoed by an all-female panel of leaders during the 2019 Trinidad and Tobago Energy Conference and Trade Show. In response to a question about the key factor for energy-sector leadership, one of the panellists, BHP's Vice President of Technology for Petroleum & Legacy Assets, Kristen Ray, said, "I think it is the humility to say we were wrong and be able to shift the direction of strategy, and not feel the need to carry it on and look bold."[4]

They Are Intrepid

All the women expressed a willingness to take risks, to try new things, to get their hands dirty. They love a good adrenaline rush and aren't put off by the possibility of short-term discomfort in the pursuit of solutions they believe in.

The career trajectories of all the women took unexpected turns, but stepping into roles and situations that made them uncomfortable at first, ultimately unearthed hidden strengths. In several cases, the sharp detours were prompted by a boss who saw something in them that they didn't yet see in themselves.

These women don't mind being pushed into the water – they are willing to doggy-paddle, to tread water, or splash about until they figure out how to swim. They refuse to drown, which brings me to my next point...

They Are "Gritty"

Grit, according to Angela Duckworth[5], who wrote a book on the topic, "is about having an ultimate goal and holding steadfast to it. Even when you fall down. Even when you screw up. Even when progress toward that goal is halting or slow." To be gritty means to pursue something with consistency of interest and effort. As Duckworth explains it, commitment to that ultimate goal renders you adaptable in relation to lower-level goals. So, if something doesn't work, gritty people don't give up, they find another way. Isn't that exactly what Marny advised? And I imagine grit is exactly the quality needed for a woman working in the hostile technical/operations side of the oil and gas industry, or a woman relocating her family halfway across the world to a new culture – technical experience and relocation being the traditional prerequisites to advancement in the industry.

They Are Good Communicators

These women innately know or have learnt:

- how to use their words to connect with and influence people;
- when to shut up and listen; and,
- what to listen for.

They Have a Support System

These women rely on supportive relationships, whether those relationships involve a spouse, a parent, a circle of girlfriends, or a professional network. Their supporters help the women to bear what was referred to in the last chapter as the "quadruple burden" – the unpaid extra work women must assume in a male-dominated industry. So, the support system may help carry the weight of family obligations or provide relief from the emotional weight of a hostile work environment or offer practical advice on how to navigate that environment.

In essence, the women I spoke with are not too big to ask for help. They cultivate, value, and are quick to acknowledge their supportive relationships. They know they cannot be successful in a vacuum.

They Practice Allyship

In addition to relying on the allies in their support networks (including mentors, coaches, and sponsors), these women made a point of serving as allies to others. Most of them said they enjoy working collaboratively and are eager to support other women. They are generous teachers and seem practically ebullient with a desire to share what they've learnt with those coming after them – as is abundantly clear from their participation in this book! In fact, I could not discern, in any of these women, tell-tale signs of unhealthy female competition in the form of "gatekeeping" or "Queen Bee Syndrome"[6].

They Like to Work

Over and over again, during these interviews, I heard, "I like to work."

I wondered: *What does that really mean?*

Different women explained it differently: I like a challenge, I like solving problems, I enjoy what I do, I like making my own money, I like going out every day and interacting with people, I don't like being idle. The bottom line for these women seems to be that a significant portion of their sense of satisfaction is connected to feeling productive, i.e., that they have covered ground, contributed toward some larger progress, and made a discernible impact on the outside world. So, maybe – STEM degree holder or not – these women embody the laws of physics which define Work as the application of Force over a Distance.

They Exhibit Emotional Intelligence[7]

Emotional intelligence is defined as the ability to understand and manage your emotions, as well as to recognise and influence the emotions of those around you. Psychologist Daniel Goleman, who popularised the term, has suggested that while intellectual and technical know-how are mere entry-level requirements for executive positions, a high degree of emotional intelligence is what sets apart the most effective leaders.

Every woman in this book declared, as one of her leadership strengths, her emotional intelligence – which is typically broken down into four core competencies:

Self-awareness:- The ability to recognise your emotions' effect, both on yourself and on your team.[8] To bring out the best in others, you first need to bring out the best in yourself.

Self-management:- The ability to manage your emotions, particularly in stressful situations, and to maintain a positive outlook despite setbacks. Leaders who lack self-management are prone to knee-jerk reactions, whereas emotionally intelligent leaders keep their reactive impulses in check while they craft reasoned responses.

Social awareness:- The ability to read the room, dude. Leaders who excel in social awareness practice empathy[9] by striving to understand their colleagues' feelings and perspectives, which enables them to communicate and collaborate more effectively.

Relationship management:- The ability to influence others and resolve conflict. It's important for a leader to promptly and adequately address thorny issues as they arise, by embracing rather than avoiding tough conversations.[10]

While emotional intelligence is obviously a benefit to anybody, in any industry, it's not hard to conceive of why it would be of particular utility to women in a male-dominated industry, who must regularly contend with the disadvantages of being tokens, many of which were mentioned by the women in this book:

- heightened visibility;
- unequal scrutiny;
- isolation and exclusion from the "ole boys'" after-work social (read "networking") scene;
- presumed incompetence;
- role encapsulation ("Oh, look: a woman! She must be the secretary."); and,
- the double bind of being impugned for their more masculine behaviours (such as asserting clear-cut authority over others), as well as for their more feminine behaviours (such as being especially supportive of others).

Knowing how to navigate all of this – what to take issue with and what to ignore, how to respond thoughtfully instead of react, when and where to escalate – is crucial to survival in the oil and gas industry, as the women's stories testify. And, perhaps most importantly (as I'll discuss later), a woman in the oil and gas industry needs emotional intelligence to know when she's being gaslighted, i.e., to glean the difference between situations where she is the problem (and can improve) and where the problem is the systemic situation.

They Recommend Forgiveness

Most of us, as humans, are driven by highly sensitive internal systems of justice. When someone wrongs us, or when we perceive we have been wronged, or when we have a hurtful experience, we begin to stew. We want vengeance, we want the perpetrator to burn in hell, we want an apology – and not just a perfunctory one, they must mean it, for Chrissake! – and nothing less will suffice.

Yet many of the women in this book, going against this darkness in our nature, advocated "forgiveness".

Psychology supports them. According to the Zeigarnik Effect, a finished task needs not to be recalled, and by forgiving those who have wronged or offended them, individuals "finish the task" or close the loop on difficult situations. By closing the loop, forgiveness reduces rumination, which makes way for reflection to occur, which in turn translates into better physical and psychological health.[11]

Rumination is the tendency to dwell on, rehash, and re-evaluate events or experiences from a negative emotional perspective. Research has shown that women are more prone to rumination than men, because we have more "social-emotional roles and stress-related problems because of the role, and therefore... are more overwhelmed by negative emotions..." According to the Response Styles Theory, women are more likely than men to feel responsible for difficult events they experience, which might escalate ruminative thoughts.[12]

So, if we accept that a woman working in the male-dominated environment of the oil and gas industry:

- will carry multiple burdens, roles, and stresses; and,
- will suffer a multitude of slights and wrongs (whether conscious or unconsciously inflicted),

then we must accept that, without something to mediate the harm, she will be susceptible to psychological and possibly even physical discomfort that may well force her out of the industry. What the women in this book are suggesting, is that forgiveness of self and others, is that mediator.

What is "forgiveness"? It's so hard to define, but we know it when we feel it – or rather, when we no longer feel the urge for vengeance, anger, and avoidance.[13]

In this professional context, forgiveness is not a kumbaya experience (let's hug it out and forgive the abusers!), but rather, a personal trait akin to resilience. It's the ability to get angry, yes, but then – instead of getting stuck in a mental rut waiting for apologies or changed behaviours from others – to let go and move forward purposefully.

Refining the Lessons

As I sat down to transmit these courageous women's stories, I felt a great sense of responsibility – and I don't mean only my professional obligation, as a writer, to portray accurately their experiences in the oil and gas industry. Beyond that, I felt a deep personal responsibility, as a woman who – like so many others both within and outside the industry – holds too many memories of "that time when..." she was diminished at work, simply for being a woman.

So, to honour the courage of the women in this book, I will make some awkward disclosures of my own. I know what it's like to be asked to make tea and coffee for everyone at the table, because I'm the only woman in the boardroom. I know what it's like to give a legal opinion, hear everyone rubbish it, then watch them turn around and applaud the genius of an old white man who shows up and charges thousands of pounds sterling to give the very same opinion. I know what it's like to leave a job for a better opportunity only to have my ex-boss handle the perceived rejection by casting aspersions on how I got the new job. (When a man gets promoted, does anyone say, "He slept his way to the top?")

There was no way I could listen to the women I interviewed without silently and secretly comparing wounds. And there was no way I – an ambitious, competitive professional with dreams of founding and becoming Managing Editor of her own publishing house – could mark and collate the women's survivor characteristics, without using the list as a yardstick against which to measure myself. Am I a continuous learner? Am I emotionally intelligent? I scored higher in some areas than others. Thankfully, I assured myself, some skills can be learnt or developed as part of leadership training – communication, conflict resolution, etc. Other skills, as well as a certain degree of self-awareness, can be elicited by a good professional coach, if you're willing to pay for the service. Building out a support system might begin with as simple a step as joining one of the professional women's networks.

But then, I also found myself taking off the professional hat and slipping into the protective motherhood role, assessing my little daughter's burgeoning personality: Is she gritty enough? Do I need to do more to help her become intrepid?

I had to stop myself. I saw that I was making the quintessential mistake of any girl or woman who has imbibed the Self-Sacrificial Kool-Aid. And haven't most of us Caribbean women taken a sip, haven't some of us watched our mothers and mother-figures chugging that Kool-Aid, and haven't their well-meaning hands held the cup to our impressionable lips and bade us drink: "Girl-child, if you do more, if you fix you – if you learn more, if you try harder, if you pray longer, if you 'band yuh belly' tighter – the unfair situation will change."?

But no, the world's unfairness will not change only by dint of our attempts to "fix" ourselves. And so, we do and do and do, and we rush in to fight fires that we didn't start and couldn't possibly put out – the classic Glass Cliff[14] situation, where a woman is put into power during times when failure is more likely – until we ourselves burn out in a flaming ball of bitterness.

So maybe this is empowerment: becoming so secure within myself that I cannot be gaslighted by anyone. So secure that I do not rush to assume responsibility for burdens that are not mine to carry. So secure that I do not dash to the edge of The Glass Cliff as soon as someone yells, "We're in crisis, bring a woman!" So secure that I am able to choose opportunities for change wisely and with the best partners, and then to endure either until change comes or until I am satisfied that I have taken a shoe heel and made a good-sized crack in that infamous sheet of glass ceiling.

And here's another thought: Is anyone else wondering what would happen if we moved away from "fixing the women" and started talking about "fixing the men"?

What I'm suggesting here is not mere male-bashing. Alienating men, as the women in this book have taught me, will never correct the imbalances in the male-dominated oil and gas industry. On the contrary, men must be a key part of the industry's evolution. As you'll recall, most of the women I interviewed credit male mentors and sponsors for helping them rise to the top – proof that secure, progressive-thinking men are instrumental to change.

So, what we need is more "good guys". But how do we find them? Is there a way for men to demonstrate their ability to overcome their ingrained biases? Now, I'm not talking about a one-day "Unconscious Bias" training at the Marriott with snacks and lunch. I'm afraid I don't hold much stock in the notion that a "baddie" with gender-bias training can truly become a "goodie".

But is there a way to weed out the "baddies"? Appears so! The answer seems to lie in the realm of psychometric testing – a well-worn recruitment tool in every major industry.

Studies have found that a certain psychometric profile – the Short Dark Triad of Personality scale – is "significantly related to both hostile and benevolent sexism". This means that employers could test job applicants for the relevant variables, identify those with a propensity for sexism, and avoid hiring them.[15]

Testing for "the dark triad of personality" may sound far-fetched. But in fact, the oil and gas industry already screens job candidates for personality and behaviour traits, such as neuroticism, conscientiousness, and situational judgment, that are likely to enhance or sabotage the work environment. Why not add testing for sexism? For an industry that boasts of its capacity for innovation, this seems a natural next step.

Would that be too much to ask, though? Would such testing so diminish the pool of male talent that we'd end up chasing leprechauns? Or are good guys plentiful in the oil and gas industry? Sure, I could just take the word of the women in this book about their fantastic male mentors, but that would be like someone telling me how extraordinary a hummingbird is, versus me actually seeing one defying gravity over a flower. So, I decided to ask around independently and try to find a senior man, unconnected with them, who had a reputation for being an ally of women. If I found such a real life, flesh-and-blood "good guy", I planned to ask him why he was so different to the oil and gas norm.

So, having started this book by talking to a man, I ended it by talking to another: Zaid Khan, an engineer of forty years' experience (UWI, 1973), a Fellow of the Association of Professional Engineers of Trinidad and Tobago, and owner of one of the premier inspection engineering companies in the region. Mr Khan is so well-regarded that he was recently called upon to testify, as an expert, before The Commission of Enquiry into the 2022 Paria tragedy which shocked the regional oil industry. I took that as indicative of his veracity. Also, as an independent businessman with a reputation so firmly entrenched, he had nothing to lose by talking straight with me. Mr Khan shared that his company, at the date of interview, employed an equal number of male (four) and female (four) engineers; that he had never viewed a woman's gender as inhibiting her performance; that, in recent years, the number of female applicants far outstripped male applicants; and that, in his opinion, both in academic qualification and on-the-job performance, women were surpassing men. Mr Khan saw himself as a proponent of equal opportunity for women in the engineering field:

"There's a psyche in Trinidad that keeps women back: *Aye! You see them heavy-duty work, women can't do that!* But in my experience, that's not true. I think we have passed that stage where men should belittle women in this industry. I give them their due respect. And the fellas in my company know that, they know they have to be respectful too."

I tested his self-characterisation, though, by speaking to a woman who had worked for him as a young inexperienced trainee. This woman, who asked to remain anonymous, recounted an incident in which, early one morning, she was scheduled to go to an offshore platform as part of an inspection engineering team, but was stopped, by the Client's staff, from boarding the boat "because she was a female." An impasse ensued. Mr Khan had to be roused from his bed. It would have been easy for him to tell his team to leave the young woman behind or to replace her – especially since she was the only female employee and he had many other qualified men who could have taken her spot. However, he said, "If they don't allow her to go, then nobody will go," and, indeed, nobody went that day. After some back and forth, it turned out to be an issue of whether the Client was satisfied that the right insurances were in place to cover the female trainee. The matter was resolved and the inspection team, including that same woman, were able to resume their job over the subsequent days. Mr Khan remembers the incident only vaguely, but the woman has never forgotten it. She said:

"It was my first time in a situation like that. I was kinda embarrassed, I felt like I was a problem and I was keeping back the work. I was mentally prepared to return home, and so Mr Khan's response was a pleasant surprise. It made me feel really respected and important as a female inspector that day. Knowing my employer would stand up for fair play gave me a sense of belonging, like I could have a future in the industry. Especially because this was my first real job after Technical School and I was coming from a home where I didn't have a father or brothers."

I'm glad this woman had a boss as forward-thinking as Zaid Khan. But Khan was no hero and I'm not suggesting that he be deified. No, he was just doing the bare minimum that any man – any leader – should be expected to do. If oil and gas must continue being male dominated for some (hopefully) limited time in the future, Khan – at least in his attitude to women's value and capability – should represent the normative man, not the exception. If psychometric testing helps the industry establish this new norm, then I'm all for it. The true heroes, the ones I want my daughter to grow up revering as role models, are the women who, in the inclement gender environment of the 20th and early 21st century oil and gas industry, battled the tide and survived to share their stories with us, in this book.

Perhaps, given that oil and gas has, for so long, been regarded as a STEM field, we should conclude with a STEM principle. Science says that energy can neither be created nor destroyed – only converted from one form to another. The oil and gas industry cannot redo or erase its gendered past any more than these women can expunge their hurtful experiences. But we can continue to flare off the bad – as we've done in this book – and convert the good, the lessons, in order to find that elusive cocktail: the most efficient and sustainable combination of male and female energies to power what looks to be, for our sons and daughters, a thrilling new world. ■

Endnotes

1 Deffler, S.A.; Leary, M.R.; Hoyle, R.H. 2016. "Knowing what you know: Intellectual humility and judgments of recognition memory". *Personality and Individual Differences* Vol. 96: 255–259. ISSN 0191-8869. https://doi.org/10.1016/j.paid.2016.03.016.

2 Note 1, *supra.*

3 Note 1, *supra.*

4 *Kaieteur News.* February 11, 2019. "Leaders' humility critical to energy sector's success – All female panel at TT Conference". https://www.kaieteurnewsonline.com/2019/02/11/leaders-humility-critical-to-energy-sectors-success-all-female-panel-at-tt-conference/

5 Research psychologist and author of *Grit: The Power of Passion and Perseverance.* Scribner, 2016. https://angeladuckworth.com/qa/

6 The phenomenon where women who are individually successful in male-dominated industries and accomplish executive positions are more likely to endorse gender stereotypes. That is, they tend to consider the women they supervise as competitors with negative attitudes towards them. See: Staines, G.; Tavris, C.; Jayaratne, T. 1974. "The Queen Bee Syndrome". *Psychology Today* 7(8): 63–66.

7 Landry, Lauren. April 3, 2019. Harvard Business School Online. "Why Emotional Intelligence Is Important In Leadership". https://online.hbs.edu/blog/post/emotional-intelligence-in-leadership

8 According to research by organisational psychologist Tasha Eurich, 95 percent of people think they're self-aware, but only 10 to 15 percent actually are, and that can pose problems for your employees. Working with colleagues who aren't self-aware can cut a team's success in half and, according to Eurich's research, lead to increased stress and decreased motivation. See: Note 7, *supra.*

9 Global leadership development firm DDI ranks empathy as the number one leadership skill, reporting that leaders who master empathy perform more than 40 percent higher in coaching, engaging others, and decision-making. See: Note 7, *supra.*

10 Research shows that every unaddressed conflict can waste about eight hours of company time in gossip and other unproductive activities, putting a drain on resources and morale. See: Note 7, *supra.*

11 Mróz, J.; Kaleta, K. 2023. "Forgive, Let Go, and Stay Well! The Relationship between Forgiveness and Physical and Mental Health in Women and Men: The Mediating Role of Self-Consciousness". *International Journal of Environmental Research and Public Health* 20(13): 6229. https://doi.org/10.3390/ijerph20136229

12 Note 11, *supra.*

13 Note 11, *supra.*

14 Bruckmüller, S.; Branscombe, N.R. 2011. "How Women End Up on the "Glass Cliff". *Harvard Business Review* 89(1-2): 26. https://hbr.org/2011/01/how-women-end-up-on-the-glass-cliff

15 Bria, P.; Etchezahar, E.; Ungaretti, J.; Gómez Yepes, T. 2022. "The dark side of sexism in Argentina: Psychometric properties of the Short Dark Triad Personality measure and its relation with ambivalent sexism". *Frontiers in Psychology* Vol. 13. doi: 10.3389/fpsyg.2022.962934

Picture Sources

Author's Photo: Damian Luk Pat	X
Paria Publishing Archives	1, 3, 5, 6, 9
After an old Shell Map	8
From an old Shell documentary	10
Website of the Office of the Prime Minister of Trinidad and Tobago, www.moc.gov.tt	11
The University of the West Indies - ICTA Documents SC 118 Box 3 Album 2	12
After an older version on www.energy.gov.tt	13
PLIPDECO	15
Private Contributor	22/23, 155
BPTT	65
PETROTRIN	97
Staatsolie Maatschappij Suriname N.V.	107, 110, 117
After a Staatsolie Map	115
Infographics in this chapter from Boston Consulting Group's *Untapped Reserves 2.0*	160
World Bank	173
AI image supplied by the author/publisher	Cover, VIII, 180

All other images were supplied by the interviewees and their companies.

www.ingramcontent.com/pod-product-compliance
Lightning Source LLC
LaVergne TN
LVHW072013150826
845684LV00009B/67

9789768291943